Coherence, Continuity, and Cohesion
Theoretical Foundations for
Document Design

Coherence, Continuity, and Cohesion
Theoretical Foundations for
Document Design

Kim Sydow Campbell
Air Force Institute of Technology, Ohio

LEA LAWRENCE ERLBAUM ASSOCIATES, PUBLISHERS
1995 Hillsdale, New Jersey   Hove, UK

Lawrence Erlbaum Associates, Inc., Publishers
365 Broadway
Hillsdale, New Jersey, 07642

Library of Congress Cataloging-in-Publication Data

Campbell, Kim Sydow.
    Coherence, continuity, and cohesion : theoretical foundations for
document design / Kim Sydow Campbell.
        p.   cm.
    Includes bibliographical references and indexes.
    ISBN-0-8058-1301-2 (c). — ISBN 0-8058-1703-4 (p)
    1. Technical writing.   I. Title.
    T11.C29   1994
    808′.0666—dc20
                                                    94-34527
                                        CIP

Books published by Lawrence Erlbaum Associates are printed on acid-free
paper, and their bindings are chosen for strength and durability.

Printed in the United States of America
10 9 8 7 6 5 4 3 2 1

To my parents, Val and Phyllis,
for everything

Contents

Acknowledgments

Most of the work that has resulted in the publication of this book was accomplished during my first two years on the faculty of the Air Force Institute of Technology (AFIT). I want to thank my department head, Professor Bob Weaver, and my current dean, Colonel Tom Schuppe, for allowing me two academic quarters during those first two years in which I was free to pursue the ideas presented here without the distractions of teaching. Thanks to my former student, Captain Joe Swillum, USAF, for performing the briefing that formed the basis of the transcription that appears in Fig. 5.1. Thanks also to Dean Lehman and Greg Smith, AFIT's graphic artists, for their work on the figures presented in this book. In addition, I want to acknowledge the professionalism of the folks at LEA: my editor (and fellow Cornhusker), Hollis Heimbouch, and production editor, Kathryn Scornavacca. Most importantly, special thanks are due my mentor and friend, Frank Parker: for commenting on parts of this manuscript; for providing Fig. 4.2; for moral support; and mostly for his integrity and intelligence; this book would not exist except for his influence and inspiration.

Introduction:
A Research Agenda

'Sense' has two senses, one perceptual and the other linguistic. We have tried to take care of them both, for we feel the two are not as different as they are sometimes made out to be.
—Miller and Johnson-Laird (1976, p. vi)

The quote above from the preface of Miller and Johnson-Laird's landmark book, *Language and Perception*, led me to formulate the theoretical core of this book—that the cognitive principles which explain why humans 'sense' unity in a succession of sounds (which therefore constitute a whole musical piece) or in a configuration of visual shapes (which therefore constitute a whole object) are the basis of principles that explain why we 'sense' unity in a string of sentences or a series of computer screens (which therefore constitute a whole text or discourse). More specifically, I will argue that one aspect of discourse coherence, continuity, is analogous to visual and auditory unity, as studied by the Gestalt school of psychologists. In addition, I argue that Gestalt principles like proximity and similarity describe how cohesion is produced through the use of the full range of discourse elements (e.g., from white space and typography to beeps and pauses to parallel syntax to synonymous lexical items and deictic terms). Thus, I believe cohesion produces continuity, one type of coherence, in discourse. More generally, then, it is my premise in this book that humans extend the use of cognitive perceptual principles like that of proximity, originally used in response to interaction with visual and auditory phenomena, to the more complex, relatively late-developing cognitive task of discourse comprehension and production.

The notion that discourse unity might somehow be analogous to auditory and visual perceptions of unity appealed to me mainly because of my practical experience as a technical communication teacher and a scientific editor. In particular, I regularly comment on the design of communication, including the auditory or visual qualities of an oral or written discourse along with its linguistic qualities. As one simple example, I often note that more attention to page layout or design would create a more clearly unified, coherent, and usable discourse (e.g., placing a diagram next to relevant prose in a technical manual). As I document later in this chapter, the practical importance of page design is well recognized by other technical communication professionals.

Although page design might appear to some to be of only trivial intellectual interest, I believe that its influence on a human being's sense of discourse unity must reflect the lawfulness of the human mind. Therefore, I conclude that, like a theory of language, a theory of discourse unity that could account for the unifying effects of visual and auditory as well as linguistic elements might be of great intellectual interest as a contribution to a theory of mind or cognition.

Unfortunately, my training as a linguist offered me no theoretical framework within which I could understand how the full range of elements (e.g., visual as well as linguistic components) are involved in creating unified texts or discourse. Not surprisingly, the unifying role of non-linguistic elements has not often been considered within linguistics and, in terms of linguistic elements, the unifying role of phonological and syntactic elements has been largely ignored in favor of semantic elements. Although I found that psycholinguistic research has indeed considered the effects of non-semantic and even non-linguistic elements, that research provided no general theory or explanation for why those elements enhanced the unity and coherence of discourse. As a consequence, I found no one theory that provided a satisfactory explanation of the relationship among these unifying linguistic and non-linguistic elements. Furthermore, no one theory provided a satisfactory explanation of the role of these elements in establishing coherence.

Therefore, my theoretical knowledge provided no way to account for my intuitive, practical experience until I read Miller and Johnson-Laird's (1976) work, which argues that a significant part of the semantic component of language is founded on perceptual concepts (e.g., the perceptual concepts of "motion" and "direction" are the semantic foundation of *ascend* [motion up], *pivot* [motion around], and *depart* [motion away]). These authors, however, concentrated on lexical or word meaning. It is my goal in this book to extend their general assumption about the relationship of perception and language to the level of discourse. The recognition of the relationship between perception and language inspired me to explore the utility of Gestalt theory (which provides perceptual principles describing how auditory and visual unity is achieved) as a theoretical foundation for understanding the role of and relationships among all discourse elements in achieving discourse unity and coherence.

The remainder of this chapter expands the preceding discussion in order to justify the need for basic research that establishes a new theory of coherence and cohesion. This basic research, in turn, may provide a theoretical foundation for further applied research involving the design of communication. First, I demonstrate the importance of document design within fields outside linguistics and document the recognition for further research in this area. Second, I demonstrate that no linguistic or psycholinguistic research has been able to account for the unifying effects of both linguistic and non-linguistic discourse elements and for the role of these elements in establishing coherence. Third, I clarify the terminology used in the remainder of this book and outline the goals and design of the following chapters.

DOCUMENT DESIGN IN TECHNICAL AND SCIENTIFIC COMMUNICATION

Shriver (1989a) defined document design as "the theory and practice of creating comprehensible, usable, and persuasive texts" (p. 316). Within the field of technical and scientific communication, the importance of document design (also called "information design") has been increasingly recognized by teachers, professionals, and researchers since the 1980s. As Benson and Burnett (1992) noted:

> An increasing number of teachers of technical, business, and professional communication are integrating information about visual design into their classrooms. In addition to these pedagogical progressives, workplace practitioners have begun to recognize the need to apply research findings about effective designs of visible language. (p. 87)

Technical communication pedagogy reflects this emphasis in textbooks of the 1990s. For instance, the title of Mathes and Stevenson's (1991) textbook is *Designing Technical Reports*, and Houp and Pearsall's (1992) textbook has one of its five core parts headed, "Document Design in Technical Writing," which is comprised of three chapters and 154 pages. As Anderson (1987) wrote in his textbook chapter, "Designing Pages," "You build your written messages out of *visual* elements. These visual elements are dark marks printed on a lighter background: words and sentences and paragraphs; drawing, graphs and tables. They are *seen* by readers before they are read and understood" (p. 448). Advanced courses in technical communication are also teaching students the importance of document design. For example, Rude's (1991) textbook offers the following advice to technical editing students:

> An editor cares about format because format is functional. Format influences how well a reader uses and understands a document. The five main functions of format ... [are] to meet reader expectations, to motivate readers, to provide access to

selected parts of the document, to aid the readers in comprehension, and to facilitate its continued use. (p. 288)

Professional technical and scientific communication practice also reflects this emphasis on document design. For instance, Benson (1985), a research associate in the Design Center of the American Institutes for Research, advised professional communicators: "[t]o design a document well, you need to imagine what linguistic and visual organizers will help readers understand how the text is structured" (p. 36). In addition, the Society for Technical Communication's professional journal, *Technical Communication*, has published special issues on document design (e.g., *Document Design Moves into the Next Decade*, edited by Shriver, 1989c).

Professional communicators have also promoted the importance of research in the area of document design. As IBM's Brooks (1991) wrote:

most [technical communicators] would probably agree that text set in all uppercase letters is harder to read than mixed-case text. That a well-designed serif type is easier to read than sans serif. But are you really sure why, or do you just *know* that? If you're challenged on a question like that, it helps to be able to back up your opinion with published research results or studies. (p. 183)

Shriver's (1989a) review of document design research in the 1980s includes the following questions as part of the agenda for the 1990s:

- What are the principles underlying the visual design of effective text? Do some visual information structures meet readers' needs better than others?
- What is the role of writers' knowledge in document design? Subject-matter knowledge? Linguistic knowledge? Perceptual knowledge? Strategic knowledge? Rhetorical knowledge?
- Which text-evaluation methods are best suited for judging text quality? . . . Can we develop more sensitive text-evaluation methods than are currently available? . . . (p. 325)

Thus, despite the obvious importance of document design within the field of technical and scientific communication, there is a perceived need for more research that illuminates the principles describing effective design and takes into account the various types of knowledge writers and readers bring to the task of communication production and comprehension. This research is important as a means of providing general principles that form a foundation for evaluating text quality in pedagogical and professional practice. General principles are crucial for providing novice communicators with the knowledge required to diagnose rather than simply detect problems with the texts they produce (Flower, Hayes, Carey, Schriver, & Stratman, 1986, p. 47).

RESEARCH ON COHERENCE AND COHESION

One logical place to look for systematic, general principles describing how unity or coherence is established in discourse is, of course, linguistic and psychological theories of coherence and cohesion. The importance of unity or connectedness as an aspect of coherence is universally recognized (e.g., as noted by Hatakeyama, Petöfi, and Sözer's, 1985, review of research on coherence in textlinguistics or by the title of Charolles and Ehrlich's, 1991, review of research on coherence, "Aspects of Textual Continuity"). To begin, I first define coherence and then take up the question of whether previous research can provide the general principles needed to answer the research questions mentioned earlier from document design.

Halliday and Hasan (1976) wrote, "[a] text is a passage of discourse which is coherent in these two regards: it is coherent with respect to the context of situation . . . ; and it is coherent with respect to itself, and therefore cohesive" (p. 23). Similarly, Hatakeyama et al. (1985) distinguished two types of textual unity or connectedness: co-textual and con-textual unity. **Coherence** (contextual unity) involves connections between the discourse and the context in which it occurs. For example, consider the following excerpt from a proposal written by a group of professional civil engineers:

Example 1.1.
Mr. Krishan Saigal, P.E., will serve as Lead Engineer. Mr. Saigal's primary tasks will include:

- Plan and provide direction for technical work elements.
- Coordinate technical direction of subcontractors.
- Assist in coordinating and disseminating project-related information to the Project Team . . . .

Mr. Saigal will also serve as Construction Manager for the Project Team, with the following primary responsibilities: (SCS Engineers, 1991, p. 2-2)

Note that, although it shows some signs of connectedness, it does not constitute a "whole" text. More specifically, although describing both of Mr. Saigal's duties consecutively (i.e., with no intervening, extraneous material) establishes some coherence, the excerpt lacks the quality of completeness: in other words, the completion of the last sentence, which should include a list of duties related to the role of Construction Manager. The reader expects completeness in a proposal like the one this excerpt comes from. Thus, coherence describes the relationship between the discourse and the context in which it occurs.

Cohesion (co-textual unity) involves connections within the discourse. As one example of cohesion, in Example 1.1, note that the word *technical* is repeated in the each of the first two enumerated items and that two morphologically similar terms, *coordinate* and *coordinating*, are repeated in the last two enumerated items, thus creating a connection among the three items. As another example, note that bullets and parallel syntax are used; each of the three items begins with a bullet and a verb in the same tense/aspect, thus creating a connection among the three items.[1] In sum, cohesion describes connections among the elements within the discourse.

Coherence and cohesion theories that have received the most attention in applied research in writing and communication include Grice's (1975) theory of implicature, Mathesius' theory of functional sentence perspective (also known as the given-new contract or thematic progression; see Daneš, 1974), and Halliday and Hasan's (1976) theory of cohesion. First, for example, Cooper (1982) used the theory of implicature to explain how writers use their knowledge of the world to communicate with their readers, and Riley (1988a) used it to provide insight into recommended strategies for communicating negative messages in business. Second, Vande Kopple (1982) used functional sentence perspective (FSP) to show that writing can be made more comprehensible by following certain patterns of information arrangement; and Thompson (1985) used it to provide strategies for improving the communication of technical writing students. Third, much applied research in writing and communication has been based on the theory of cohesion developed by Halliday and Hasan (1976) in *Cohesion in English*. For example, Johns (1980) used their theory to investigate cohesion in business discourse; Witte and Faigley (1981) used it to investigate the relationship of cohesion, coherence, and writing quality in the written products of freshman English students; and Myers (1991) used the theory to investigate the relationship between cohesion and subject-matter knowledge of readers.

Unfortunately, however, this applied research has noted some problems with coherence and cohesion theory. First, the theory of cohesion "requires further change . . . [because] most of the Halliday and Hasan coding was done on British literature, especially *Alice in Wonderland*; . . . items which appear in Lewis Carroll's writing are not those typical of modern business writing" (Johns, 1980, p. 41). If we consider technical manuals, then the applicability of the cohesive devices in Carroll's writing become even less representative. For instance, Walter (1992) noted that technical orders (product manuals used by the United States Air Force) average 40% graphic content and 20% tables, with only the remaining 40% consisting of prose (pp. 13–17). However, Halliday and Hasan chose to limit their theory of discourse unity to only semantic elements. The authors noted that they were "excluding from consideration the effects of formal devices such as syntactic parallelism, metre and rhyme . . ." (Halliday & Hasan, 1976, p. 10). In addition, they omitted FSP from their categories of cohesion because it is structurally produced.

---

[1]Because these connections do not involve semantic elements, these unifying elements are not included in Halliday and Hasan's theory. I explore the consequences of this later in this section.

As I wrote in Campbell (1991):

> While choosing to limit the scope of *Cohesion in English* to semantic cohesion is in itself fairly unremarkable, the fact that analysts of written texts have continued for fifteen years [seventeen at the time of this writing] to use Halliday and Hasan's theory of cohesion without substantial addition is remarkable. (p. 223)

I also noted that there are only a few studies that have expanded the original theory of cohesion. For instance, Stotsky (1983) argued for a modification of one of the original categories of semantic cohesion; Hartnett (1986) distinguished between two functions of semantic cohesion; and Markels (1983) investigated syntactic cohesion. However, only Markels introduced another category of elements and then only one of many possible categories. In response to this lack of expansion in cohesion theory, I attempted to establish the range of non-semantic cohesive elements by analyzing technical discourse: FSP, syntactic parallelism, and graphic devices (including typography, enumeration, and chart types; Campbell, 1991). However, that work did not provide a comprehensive, theoretical framework within which the unifying effects of both semantic and non-semantic discourse elements could be understood.

Research in psychology (defined broadly enough to include psycholinguistics, educational psychology, cognitive psychology, and reading comprehension) has in fact recognized the impact of non-semantic elements on discourse coherence (e.g., see collections edited by Tzeng & Singer, 1981; and Besner & Humphreys, 1991). As one example, in 1985 Kieras wrote:

> the cognitive psychology of comprehension has tended to ignore surface structure in favor of semantic content. However, it seems clear that surface structure is normally chosen by the writer in an attempt to convey a desired meaning most efficiently. . . . Hence an adequate theory of comprehension must explain not only how readers derive the semantic content of sentences and relate them to already known information, but also how the surface form of the input is used to guide or streamline this process. (p. 103)

No doubt this recognition stems from the fact that many researchers turned from studying narratives to technical and scientific discourse in the 1980s. Kieras (1985) also made a convincing case for the social importance of such studies, noting that this is the discourse of most textbooks and also of the overwhelming number of technical documents that accompany the products of modern technology (p. 90).

Unfortunately, however, the experimental research that has investigated the unifying effect of such elements has not provided a theoretical framework explaining the relationship between semantic and non-semantic cohesive elements, nor has it clearly established the role of these unifying elements in establishing coherence. For instance, Mayer (1985) provided "some suggestions for how to increase the understandability of science text [which] are offered as 'good guesses' based on a general interpretation of our research, [and] should be subjected to additional testing" (p. 84). To illustrate, consider a few of Mayer's (1985) suggestions:

- Signal the major explanative ideas in the text such as using numbers. (For example, 'First, a pulse is sent out . . .'
- Use headings and indentations to indicate the major ideas. (For example, each idea is present on its own line.)
- Include repetition of important ideas in various wordings; build redundancy into the passage . . . (pp. 84–85)

Mayer argued that these suggestions provide writers with ways in which they can focus the reader's attention on the salient ideas in an explanatory text. Although I do not question the accuracy of Mayer's claim or the validity of his suggestions, I am not satisfied that such experimental results have explained **why** enumeration, headings, page arrangement, and semantic repetitions create more unified and coherent discourse.

A number of research studies that have applied Halliday and Hasan's theory of cohesion have provided conflicting or unclear results in terms of the relationship between semantic cohesive elements and coherence in a discourse. For instance, although McCulley (1985) found a correlation between some types of semantic cohesion and coherence, Witte and Faigley (1981) found no correlation between writing quality and semantic cohesion, and Tierney and Mosenthal (1983) found so many semantic cohesive elements that they judged cohesion analysis useless in determining coherence. In addition, some dissatisfaction with cohesion theory has involved the perceived lack of utility of quantitative analysis. For example, Hendricks (1988) wrote,

It takes Halliday and Hasan about seven pages to explain their scheme for coding the types of cohesion. . . . And when one imagines the whole text of, say, *Alice in Wonderland* subjected to such an analysis, the result is bound to be a mass of data so overwhelming as to be practically useless. (p. 104)

In sum, I believe previous research in both linguistics and psychology suggests the need for a theory of coherence and cohesion that addresses the following questions:

A. How can we account for the unifying effect of the full range of discourse elements: semantic elements, other linguistic elements, and non-linguistic elements? What is the relationship between semantic elements and other cohesive elements?

B. What role do cohesive discourse elements play in establishing coherence? Can we predict when they will and will not enhance coherence?

The main point here is that the answers to such questions precede the possibility of using theories of coherence and cohesion to answer research questions in document design. To review, I noted a perceived need within the field of technical and scientific communication for research that provides general principles that would form a foundation for evaluating document quality in pedagogical and professional practice by answering the following questions based on Shriver (1989a):

1. What are the various types of knowledge writers and readers bring to the task of communication production and comprehension?
2. What principles describe effective document design?
3. How can we develop effective methods of evaluating text quality?

A theory of discourse coherence and cohesion that answered Questions A and B would provide a theoretical framework within which Questions 1, 2, and 3 could be approached.

THE GESTALT CONCEPT OF UNITY

In order to answer the research questions listed in the preceding section, this book explores the analogy between our sense of auditory or visual unity and our sense of discourse unity. Gestalt psychologists attempted to delineate the psychological principles that would explain why humans experience visual and auditory phenomena as wholes. As one of Gestalt psychology's founding fathers, Max Wertheimer (1938), explained:

> one sees a series of discontinuous dots upon a homogeneous ground not as a sum of dots, but as figures. Even though there may here be greater latitude of possible arrangements, the dots usually combine in some "spontaneous", "natural" articulation—and any other arrangement, even if it can be achieved, is artificial and difficult to maintain. (pp. 71–72)

He used Fig. 1.1 to illustrate. Note that the most natural way of perceiving this figure is as three groups of two dots. However, it is also theoretically possible to perceive this figure as two groups of three dots. As Wertheimer (1938) argued, "it is for most people impossible to see the whole series simultaneously in the latter grouping" (p. 72).

It was the premise of Gestalt psychologists that such perceptions of unity were predictable or rule-governed, and they developed a set of theoretical principles or rules that accurately describe human perceptual predispositions. In

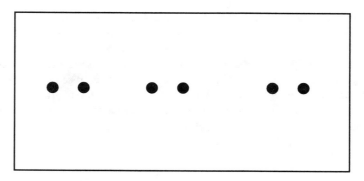

FIG. 1.1.   A Gestalt figure demonstrating the principled nature of visual perception.

the case of Fig. 1.1, it is the principle of proximity that describes our preference for perceiving groups of two dots; in other words, because the sets of two dots are physically close to each other and physically more distant from any third dot, we perceive the sets of two dots as three unified wholes.

As I noted earlier, it is the premise of this book that humans extend the use of cognitive perceptual principles like that of proximity, originally used in response to interaction with visual and auditory phenomena, to the more complex, relatively late-developing cognitive task of discourse comprehension and production. As an illustrative, introductory example, consider Example 1.1 again. Note that, despite the excerpt's lack of completeness, I sense that the excerpt is comprised of two hierarchically equal segments: one related to a "Lead Engineer" and one related to a "Construction Manager." Note also that these two segments of the excerpt are physically separated by more white space than any other parts of the excerpt. Thus the principle of proximity provides **one** explanation for our perception of two unified segments within Example 1.1. (Of course, a number of other discourse elements and principles are also involved in this perception.)

Interestingly, one of the most gifted discourse analysts, Roman Jakobson, suggested the potential importance of psychological principles like those the Gestalt psychologists called the principles of proximity and similarity. As Jakobson explained:

> The perception of similarities and contiguities ... united by parallelism leads automatically to the need to find an answer to the unconscious questions: What links the two lines? Is it an association by similarity or by contrast? Or is it an association through contiguity, and, if so, is it a contiguity in time or space? (Jakobson & Pomorska, 1983, p. 103)

A few technical communication researchers have argued for the importance of Gestalt principles as a means of describing effective visual design. For instance,

Barton and Barton (1985) found "that the treatment of visuals often consists of an *ad hoc* series of guidelines whose rationale, or theoretical basis, remains obscure" (p. 129). Thus, they suggested that Gestalt principles might be used to provide such a theoretical basis. Bernhardt (1986) followed through on this suggestion by applying Gestalt principles to an analysis of the visual design of a technical text. Similarly, Moore and Fitz (1993) applied an even broader range of Gestalt principles to a range of technical texts. Fortunately, this line of research has provided a cogent theoretical basis for understanding the visual aspect of document design. However, no research has provided a theoretical basis for understanding the role of the complete range of elements involved in document design.

Based on such Gestalt principles as proximity and similarity, this book proposes a theory of coherence and cohesion comprised of a set of principles for describing the unifying effects of the full range of discourse elements: from visual to semantic. In addition, these Gestalt principles clarify the role of these unifying elements in establishing coherence. Based on this theory, the book also provides an approach to the research questions already noted above in the field of document design.

My aim in this book, then, is to develop a theory. But I want to make clear from the outset how I view the status of that theory. The quote that follows, taken from Lerdahl and Jackendoff (an insightful study of unity in tonal music based on Gestalt perceptual principles), clarifies my view. The authors noted that the principles they discuss are conceived of as:

> [an] empirically verifiable or falsifiable description of some aspect of musical organization, potentially to be tested against all available evidence from contrived examples, from the existing literature of tonal music, or from laboratory experiments. ... We consider this book a progress report in an ongoing program of research, rather than a pristine whole. ... We feel, however, that we have gone far enough to be able to present a coherent and convincing overall view. (Lerdahl & Jackendoff, 1983, p. xii)

My own view of the theory presented in this book is quite similar. In other words, I do not intend to provide the definitive and comprehensive set of coherence and cohesion principles in this book. And, although the utility of the principles are subject to some testing throughout the book as specific discourses are analyzed, I intend for this to be viewed as a springboard for more research in coherence, cohesion, and document design based on cognitive perceptual principles. In addition, my readers should know from the outset that, of the three aspects of research on discourse coherence—structure, world knowledge, and process (Britton & Black, 1985, p. 6)—I have limited the scope of the theory developed here in order to concentrate only on the first two aspects; thus, **the process** of comprehending discourse is largely ignored.

TERMINOLOGY AND OUTLINE

Because of the lack of standard terminology within both linguistics and document design (not to mention between them), I want to establish the technical vocabulary that is used throughout this book and briefly explain those choices. In order to achieve the broadest possible theoretical coverage, I use the term *discourse* rather than *text* in order to designate communication that may be oral or written (or, for that matter, electronic) and of any length (i.e., communicative intent rather than length is the defining feature). For the same reasons, I use the term *recipient* to refer to readers and listeners and the term *producer* to refer to writers and speakers.

A *coherent* discourse is one in which a recipient perceives continuity as well as adequacy, accuracy, and clarity (see chapter 2). A *cohesive* discourse is one in which a producer has established continuity through the use of similar and proximate discourse elements (see chapter 3). The term *discourse element* is used to designate any of the full range of components that appear in communication (see Table 1.1). Let's consider a few examples from this table. Note that Table 1.1 divides discourse elements into two major categories: **non-linguistic** and **linguistic**. Furthermore, there are two sub-categories of non-linguistic discourse elements: *visual* and *auditory*. In addition, the table provides a representative sample of a number of types of visual elements. For instance, one visual discourse element is drawing (e.g., ☝ [icon of a pointing index finger]). Another visual discourse element is body gesture (e.g., an actual pointing index finger). Despite the fact that drawing can occur only in written discourse and body gesture only in oral discourse, they are both types of visual discourse elements.

Table 1.1 also provides a representative sample of auditory elements, the other type of non-linguistic discourse element. For example, one auditory discourse element is pitch. Interestingly, pitch can be relevant to both oral discourse (e.g., a deep human voice) and electronic discourse (e.g., a high-pitched, machine-made beep). Again, despite the medium of communication, pitch is a type of auditory discourse element.

Note that Table 1.1 also divides linguistic elements into types: *phonological*, *morpho/syntactic*, and *semantic* elements. For instance, stress (e.g., the louder, higher pitched vocalization of /el/ when pronouncing *boxelder*) is presented as a type of phonological discourse element. Of course, we might also recognize stress in an electronic discourse (e.g., the louder, higher pitch of three beeps). Thus, the categorization of stress as a linguistic rather than a non-linguistic component may be somewhat arbitrary here and is based on the tradition of analyzing stress within the field of linguistics. Fortunately, the categorization itself is not crucial to the theory developed here. Instead, it is important to recognize only that stress is a type of discourse element.

As a final example, note that Table 1.1 provides a representative sample of semantic discourse elements. For instance, one type of semantic discourse element

TABLE 1.1
List of Elements That May Be Involved in Creation of Discourse Unity,
Along With Representative Examples of Each Type

NONLINGUISTIC

VISUAL	example	AUDITORY	example
Typography	italic print	Pitch	deep tone
Geometric Shape	rectangular bar in graph	Rhythm	staccato
Picture/Drawing	photograph		
Color	blue background		
Grapheme	β (Greek "beta")		
Body Gesture	"ok" hand sign		

LINGUISTIC

PHONOLOGICAL	example	MORPHO/ SYNTACTIC	example	SEMANTIC	example
Initial Segment	/b/ in "box"	Voice	money "was stolen" (passive)	Synonymy	"cat" & "feline"
Coda	/ox/ in "box"	Tense/Aspect	money "has disappeared" (past perfect)	Overlap	"cat" & "kitten"
Stress	/el/ in "boxelder"	Gender/Number	"she" quit (feminine, singular)	Hyponymy	"cat" & "animal"
		Phase/Clause Structure	"in the garden" (prepositional phrase)	Antonymy	"animal," "mineral" & "vegetable"
				Deixis	"one" never knows (referent)
				Case	"Mary" stole the jewels (agent)
				Information Unit	Mary stole "the jewels" (given information)
				Proposition	Mary didn't steal the jewels = not (STEAL (Mary, jewels))

is listed as case role (e.g., in *The vet examined the horse with radiography* the three noun phrases serve different roles in relation to the verb: *the vet* is the **agent** of "examining;" *the horse* is the **patient** of "examining;" and *radiography* is the **instrument** of "examining). Again, the controversy surrounding whether case roles are truly semantic linguistic elements (rather than pragmatic elements) is irrelevant to our purpose here. Therefore, agents, patients, and instruments are considered components of discourse and are categorized as semantic for convenience. In brief, then, Table 1.1 provides a representative sample of the full range of elements that occur in discourse and that are all of interest in terms of the theory of coherence and cohesion presented in the next few chapters. A comprehensive description of such elements is outside the scope of this book, although of great potential interest within document design research.

This book is divided into six chapters. Chapter 1 has attempted to establish the importance and perceived need for a theory of coherence and cohesion that takes into account the full range of discourse elements and that clarifies the relationship between coherence and cohesion.

Chapter 2 will focus on discourse unity from the perspective of the recipient. I will establish a theory of coherence founded on four principles: relation (i.e., the perceived relevance of successive bits of information), manner (i.e., the perceived clarity with which information is given), quantity (i.e., the perceived adequacy of the amount of information), and quality (i.e., the perceived accuracy of information). In addition, I will consider the range of knowledge that influences a recipient's perception of coherence in discourse. More importantly, I will demonstrate that the principle of relation (which describes why recipients perceive coherence in discourse) is analogous to the Gestalt principle of continuity (which describes why humans perceive unity or wholeness in visual and auditory phenomena).

Chapter 3 will focus on discourse unity from the perspective of the producer. I establish a theory of cohesion founded on two Gestalt principles of continuity: similarity and proximity. I document the applicability of these principles to the full range of discourse elements: visual, auditory, phonological, morphosyntactic, and semantic.

Chapter 4 will outline the relationship between individual cohesive elements that establish global cohesion, resulting in the perception of organization within a discourse. I discuss two metaprinciples, reinforcement and conflict, which describe that interaction. In addition, I establish two global cohesion principles: intensity, and size and symmetry.

Chapters 5 and 6 will turn from theory to application and practice within document design. Chapter 5 will explore methods for applying cohesion analysis in future research. Specifically, I will address two serious complaints about past research that incorporates cohesion analysis: the questionable meaning of quantitative measures of cohesion as an indication of coherence and the requirement to exhaustively analyze every cohesive element in a discourse under investigation. Chapter 6 will discuss the potential implications of this theory for the practice and pedagogy of technical and scientific communication.

CHAPTER 2

Coherence:
The Recipient's Perspective

In this chapter, I consider coherence from the perspective of the discourse recipient. I outline a theory of discourse coherence that is suggested by the work of H. Paul Grice because his work focuses on the expectations that influence recipients' interpretation of the meaning of discourse. Interestingly, one of the coherence principles based on Grice's work appears to have a direct correlate in Gestalt theories of unity in visual and auditory perception. I demonstrate that Grice's concept of relation is analogous to Gestalt's concept of continuity. Thus, I claim that relation describes a discourse recipient's sense of one aspect of discourse coherence—continuity. In addition, I explore the gradable and variable nature of coherence, including the range of knowledge types that affect recipients' sense of coherence.

GRICE'S COOPERATIVE PRINCIPLE:
DISCOURSE CONTINUITY

Grice's (1975) William James lectures, delivered at Harvard in 1967, probably provide the most comprehensive framework for discussing discourse coherence. Grice, a philosopher, was interested in the semantics of language and, more specifically, in how implicit meaning is conveyed. His *cooperative principle* recognizes the cooperative foundation of communication by stating that discourse participants expect all contributions to be made as required based on the purpose of their interaction (Grice, 1975, p. 45). In other words, Grice's work outlines the expectations of discourse participants. Because expectations have been seen as central to the phenomenon of coherence (e.g., Witte & Faigley, 1981, p. 200;

15

Fahnestock, 1983, p. 416), Grice's work has provided a foundation for a number of research studies investigating discourse coherence from the recipient's perspective (e.g., Cooper, 1982; Lovejoy, 1987).

Grice's work suggests a categorization of the four ways in which discourse contributions may be (or appear to be) incoherent to a discourse recipient. These include: (a) **relation**—a contribution may appear to be irrelevant to earlier contributions; (b) **quantity**—a contribution may appear to offer too little or too much information; (c) **manner**—a contribution may appear to be too obscure or indirect; and (d) **quality**—a contribution may appear to be inaccurate. The first of these categories is discussed in the following section, and the other three are addressed later in this chapter.

Relation and Continuity

Grice proposed that discourse recipients expect successive contributions to be relevant. As an introductory example, consider the following discourse segment:

Example 2.1.
Data were gathered in the soil bins at the National Soil Dynamics Laboratory in Auburn, Alabama. The same soils were used as were used in the previous study: Norfolk sandy loam and Decatur clay loam. (Schafer, Evans, & Johnson, 1990, p. 779)

In contrast, consider the brief discourse below adapted from Clark and Haviland (1977):

Example 2.2.
The haystack was important. The cloth had ripped. (p. 33)

No doubt most of my readers find Example 2.1 more coherent than Example 2.2. Grice's theory suggests that this difference in coherence is due to a recipient's judgment that the information in the second sentence in Example 2.1 appears to be relevant to the information in the preceding sentence, whereas the information in the second sentence in Example 2.2 does not.

Interestingly, the Gestalt principle of continuity appears to describe an analogous preference in terms of perceiving visual phenomena. As the Gestalt psychologist, Wertheimer (1938) explained,

Additions to an incomplete object (e.g., the segment of a curve) may proceed in a direction opposed to that of the original, or they may *carry on* the principle 'logically demanded' by the original. It is in the latter case that 'unity' will result. (p. 83)

He used the visual example in Fig. 2.1 to illustrate his point. Note that the most natural way of perceiving Fig. 2.1 is as two lines intersecting at point X, one defined by points AXD and another by points BXC. Interestingly, however, it is also

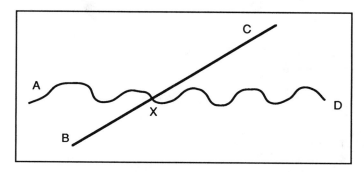

FIG. 2.1.   A Gestalt figure illustrating the principle of continuity.

possible to perceive this figure as two lines defined by points AXC and BXD. Wertheimer argued that it is the psychological principle of **continuity** that explains our preference for the first arrangement. Specifically, line XD is most naturally perceived as the logical continuation of line AX; likewise, line XC is most naturally seen as the continuation of line BX. Thus, AXD is perceived as unified, whereas AXC is not.

This human preference for sensing unity where there is visual continuity appears to be analogous to that of sensing coherence where there is discourse relation. Note, in fact, that discourse Example 2.1 might actually be described by a recipient as more unified than Example 2.2. That is, the discourse recipient's perception of coherence in Example 2.1 is dependent on the fact that the propositional content of the second sentence appears to be connected to the propositional content of the first. Thus, Example 2.1 is perceived as unified. In contrast, the recipient of Example 2.2 does not perceive the connection between the proposition expressed in the second sentence and that expressed in the first. Thus, Example 2.2 is not perceived as unified.

It is, of course, widely accepted that part of a recipient's judgment about the coherence of a discourse relies on the real-time cognitive processing during the interaction with that text (e.g., Kintsch & van Dijk, 1978, p. 389). In other words, discourse recipients experience both oral and written discourse as temporally ordered elements. It is also well established practice to refer to the "unity" of discourse among those who must analyze and/or evaluate texts (e.g. theoretical textlinguists, writing teachers, professional writers and editors, managers, etc.). Therefore, I argue that the phenomena of continuity and coherence are related (though not synonymous) and further explore the nature of that relationship later in this chapter.

Bridging: Explicit Versus Implicit Discourse Continuity

Interestingly, Example 2.2 is perceived as unified and coherent when the recipient knows the unstated discourse topic—parachuting—because this knowledge creates a logical continuity between the propositions expressed in the two sentences.

(a)

(b)

(c)

FIG. 2.2.   A musical example demonstrating the gradable nature of unity involving auditory phenomena.

Building on the work of Grice, Clark, and Haviland (1977) called the unstated or inexplicit knowledge (like that of "parachuting" in the discourse given earlier) a bridge because this type of knowledge allows recipients to connect successive propositions in a discourse into a unified and coherent whole.

In Example 2.1, then, recipients sense continuity because of the explicit connection between the successive sentences. In other words, the propositions expressed by the two sentences are explicitly related. In contrast, recipients of Example 2.2 sense continuity when they know the topic because, though there is no explicit connection, they can construct an implicit connection or bridge between the propositions expressed. Furthermore, recipients who do not know the topic find it difficult or impossible to construct a bridge, and this lack of either explicit or implicit continuity between the propositions of the two sentences makes the discourse disunified and incoherent for them.

The Gradable and Variable Nature of Discourse Continuity

It is clear that visual and auditory unity or continuity is gradable or relative rather than absolute. As Wertheimer (1938) explained, "certain arrangements are stronger than others, and seem to 'triumph'; intermediate arrangements are less distinctive, more equivocal" (pp. 82–83). Lerdahl and Jackendoff (1983) used examples similar to those in Fig. 2.2 in order to illustrate the gradable nature of unity involving auditory phenomena. Those who read music will recognize here that the five sounds represented in Fig. 2.2a clearly constitute two groups: a unit of three sounds or notes followed by a unit of two. Note, however, that the strength of this perception is less strong with the notes in Fig. 2.2b, and even less so with the notes in Fig. 2.2c because of the change in pitch of the third note. I explore the principle that explains this variability in chapter 3. For now, the point is that these examples demonstrate that we perceive relative degrees of auditory unity.

As we might suspect, discourse continuity is also a relative phenomenon. As an example, consider the following discourse variations adapted from Riley's (1993) study of ethical language use:

Example 2.3a.
News reporter: Do you think Anita Hill is lying?
Senator: Yes.

Example 2.3b.
News reporter: Do you think Anita Hill is lying?
Senator: If someone sexually harassed me, I wouldn't follow them to another job or call them repeatedly on the phone.

Example 2.3c.
News reporter: Do you think Anita Hill is lying?
Senator: I love to eat crawfish bisque. (p. 181)

Note that Example 2.3a is the most unified of the three; the senator's response (i.e., a direct answer) is an explicit continuation of the reporter's question.

In contrast, Example 2.3b is less unified than Example 2.3a. Specifically, the senator's response cannot be perceived as an explicit continuation of the reporter's question. Those of us who find this discourse relatively unified have constructed a bridge between the two discourse contributions. Riley (1993) provided a likely line of reasoning for constructing that bridge:

- The Senator has stated that he would not follow a harasser to another job nor call the harasser on the phone.
- The Senator believes that his pattern of behavior would hold for other normal humans.
- Anita Hill has testified to following Clarence Thomas to another job and to calling Clarence Thomas on the phone.
- The Senator is known to believe that Anita Hill was not, in fact, harassed by Clarence Thomas. (Note that the Senator's response would be quite discordant if uttered by a *detractor* of Clarence Thomas.)
- The Senator believes that Anita Hill lied about the harassment. (p. 181)

Note that with this line of reasoning, the recipient has formed a connection or bridge between the contribution of the reporter and senator that prompts a sense that the senator's response is a continuation of the reporter's question, thus unifying the discourse.

Note also that the bridge constructed is completely the fabrication of the recipient based on her own knowledge. It is entirely possible that another recipient might sense continuity by constructing a different bridge or by constructing the same bridge with different knowledge. In actuality, I would be willing to bet that most of my readers used highly similar knowledge to construct the same bridge as Riley did because we probably share highly similar backgrounds (i.e., in terms of education, occupation, socioeconomic class, etc.).

Example 2.3c is the least unified of the three; the senator's response is less clearly a continuation of the reporter's question than his response in Example 2.3b. Note, however, that it is still entirely possible to use our knowledge of senators and reporters to construct a bridge that unifies the discourse (e.g., the senator is avoiding the reporter's question completely). We appear to be capable of constructing a bridge between almost any two successive discourse contributions when highly enough motivated. As Clark and Haviland explained (1977):

> There are endless ways of building bridges. Yet most listeners hearing a sequence of sentences will settle on one of a small number of possible bridges. . . . What this suggests is that the listener goes about finding the intended bridge in an orderly way. He follows a set strategy he holds in common with other speakers of English. Our guess is that his main goal is to find the most direct bridge to the previous context, assuming no more than he need assume. (p. 20)

The fact that meaning is constructed by each individual recipient is certainly a well recognized point. For example, the work of Jacques Derrida is founded on the phenomenon of "misunderstanding" in reading texts. As he stated in an interview, "Each time you read a text . . . there is some misunderstanding, but I know of no way to avoid this. . . . It's something that is motivated by some interest and some understanding" (Derrida & Olson, 1991, p. 140). Derrida's very statement and the fact that he continues to attempt to convey his own ideas through discourse certainly suggests that he believes that understanding is also a fact. As Lerdahl and Jackendoff (1983) argued regarding musical "interpretation":

> Rarely do two people hear a given piece in precisely the same way or with the same degree of richness. Nonetheless, there is normally considerable agreement on what are the most natural ways to hear a piece. A theory of a musical idiom should be concerned above all with those musical judgments for which there is substantial interpersonal agreement. But it also should characterize situations in which there are alternative interpretations, and it should have the scope to permit discussion of the relative merits of variant readings. (p. 3)

Perhaps the most convincing illustration of the limits on a recipient's interpretation of discourse is provided by Britton and Black (1985). If we consider a story of 100 utterances that generates just 20 inferences per utterance (the average of potential inferences per utterance is about 750 [Rieger, 1975]), we must recognize 20^{100} (or 10^{130}) possible interpretations. As the authors noted, if recipients considered all of these possible inferences, "understanding this simple story could not have been finished by today if the universe were a dedicated computer devoted to this story since the beginning of time" (Britton & Black, 1985, pp. 4–5).

Thus, despite the potential number of inferences or bridges that a recipient might build to establish implicit coherence, the actual number is clearly constrained. My aim in this book is to construct a theory that captures those recipient judgments "for which there is substantial interpersonal agreement . . ." and that will "characterize situations in which there are alternative interpretations . . ." about coherence. I return to this issue in chapter 3, but the most important point here is that the harder we must work to construct a bridge within a discourse, the less unified and coherent we judge that discourse.

It seems obvious, then, that the knowledge and motivations of discourse participants affect their sense of discourse continuity. Interestingly, Gestalt psychologists also noted the influence of motivation (i.e., needs and attitudes) on perceptions of visual phenomena. For instance, Henle (1961) noted that "a need or attitude may operate as a vector, pointing in one direction rather than another" (p. 176). For example, a camouflaged item may be found only by subjects who are looking for it. As another example, subjects with different attitudes might both perceive animals in a particular Rorschach test card, but each might attribute different actions to those animals. Henle (1961) also noted the vital role of past experience in establishing norms for perceiving visual phenomena (p. 176). For example, consider Fig. 2.3. The most natural way of perceiving Fig. 2.3 is as two overlapping, visual wholes: a rectangle (A) and an ellipse (B). Note, however, that this figure can also be seen in other ways suggested in Fig. 2.4 (i.e., three geometric shapes labeled as A, B, and C). It is our partly our past experience with visual shapes such as rectangles and ellipses that prompts the first and most natural perception of this figure.

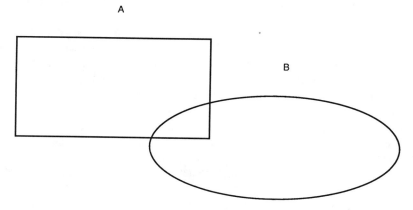

FIG. 2.3. A Gestalt figure illustrating the effect of prior experience or knowledge on visual perceptions of continuity. Perception of two, continuous visual objects.

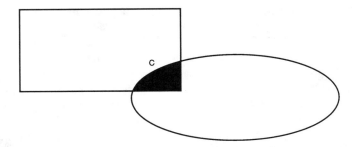

FIG. 2.4. A Gestalt figure illustrating a less natural perception of Fig. 2.3. Perception of three, continuous visual objects.

Before we move on, let me summarize the most important points made in this section. First, discourse coherence is partly a result of recipients' sense of relation between successive parts of a discourse. Second, discourse continuity established through a sense of relation between successive contributions is analogous to visual unity established through a sense of continuity. Third, continuity in both visual and discourse phenomena is gradable; we perceive relative degrees rather than their absolute presence and absence. Fourth, continuity in both visual and discourse phenomena is variable; we each bring unique knowledge, experience, and motivation to bear on our sense of continuity. In the next section, I more fully explore these types of recipient knowledge.

Recipient Knowledge and Discourse Continuity

I noted earlier that a sense of discourse continuity is variable depending on the knowledge, experience, and motivation of the recipient. Despite the presence of this variability, the fact of human communication remains incontrovertible. Communication is possible because we bring some innate psychological and some culturally learned or socially constructed knowledge that is shared with other discourse participants. It is my goal in this book to demonstrate the extent to which our judgments of discourse continuity and coherence are predictable or rule-governed (i.e., not idiosyncratic). For example, I have attempted to demonstrate that humans have an apparently innate cognitive preference for sensing unity and coherence where we see continuity. As an innate preference, the principle of continuity describes knowledge that all human discourse recipients share, thus creating some opportunity for coherence and communication.

In the remainder of this chapter and the next, I discuss other cognitive concepts that influence the sense of coherence for all discourse participants. In addition, the following section outlines the range of knowledge that discourse recipients bring to bear on questions of coherence, including those types that, by definition, are shared by many discourse participants.

In all of the example discourses so far, continuity and coherence has been viewed as a phenomenon involving a recipient's sense of the continuity of successive propositional meaning in the discourse. In other words, discourse continuity has depended on a recipient's knowledge of the semantics of English. It is important to recognize that recipients bring knowledge of domains other than English semantics to bear on their perceptions of continuity. These domains include other linguistic (non-semantic) knowledge, pragmatic discourse knowledge, other socially constructed (non-discourse) knowledge, and idiosyncratic knowledge. The following discussion considers examples of each.

Linguistic (Non-semantic) Knowledge

Aside from the influence of semantics, other linguistic knowledge may affect a recipient's sense of discourse continuity. For instance, knowledge of phonology, morphology, or syntax might also be involved. As one example, consider Examples 2.4a and 2.4b, in which the question mark signifies rising intonation and the period signifies falling intonation:

Example 2.4a.
Kevin: Are you going?
Harry: I'm going.

Example 2.4b.
Kevin: Are you going?
Harry: I'm going?

Note that the version of the discourse in Example 2.4a is unified: Harry's contribution (i.e., a statement signified by falling intonation) appears to be the continuation of Kevin's earlier contribution (i.e., a question signified by rising intonation). In contrast, Example 2.4b is not a unified discourse precisely because Harry's contribution (i.e., a question) is not the expected continuation of Kevin's. In these examples, the propositional content of both versions is identical. It is only the prosody (i.e., a phonological element) that is different. Thus, a recipient calls on all of his or her linguistic knowledge, phonological as well as semantic, to judge the continuity of discourse.

Pragmatic Discourse Knowledge

Prosody is one clue to the speech act performed by a discourse participant's contribution in an oral discourse such as Examples 2.4a and 2.4b; speech act theory describes some of the pragmatic discourse knowledge that influences recipients' sense of discourse continuity. Speech act theory is the result of work by the ordinary language philosophers, J. L. Austin (1962) and J. R. Searle (e.g., 1969, 1975, 1976). A speech act is simply the act (e.g., stating, requesting, promising, etc.) performed when a discourse participant uses an utterance. For instance, in Example 2.4a,

Kevin's utterance performs the act of questioning, whereas Harry's performs the act of stating.

Research has shown that the ordering of successive speech acts is predictable in certain cases. For example, any contribution following a directive (i.e., an order, request, or suggestion) is interpreted as a response to that directive (Davidson, 1984, p. 102). To illustrate, consider the responses of a secretary to the request from her boss who says, *Please get me some coffee*:

Example 2.5a.
OK.
Example 2.5b.
I'll make some right away.
Example 2.5c.
Do you want cream and sugar?

Example 2.6a.
No.
Example 2.6b.
Nice day, isn't it?
Example 2.6c.
It's on the table.

Example 2.7.
(silence). (Campbell, 1990, pp. 361–362)

Note that Examples 2.5a–2.5c signal compliance with the boss's request, whereas Examples 2.6a–2.6c signal refusal. Examples 2.5a and 2.6a establish discourse continuity explicitly (i.e., as expressions of compliance or refusal, they are probably the most relevant contributions possible). In contrast, the variety of propositional content conveyed by the rest of the responses require bridging in order to establish that continuity. Interestingly, even the silence of Example 2.7 is interpreted as a response of either compliance (if the secretary gets the coffee) or refusal (if he or she does not) to the boss's request. Thus, a recipient's sense of continuity in any of the discourses just given involving the nonexplicit responses is dependent on his or her knowledge of speech act ordering, a kind of pragmatic discourse knowledge.

Like directives, any contribution following a question is interpreted as an answer to that question. It is this knowledge of the ordering of speech acts in the case of Example 2.4b that has probably caused many readers to attempt to see Harry's contribution as an answer despite the discontinuity of his rising intonation.

So far we have seen that perceptions of continuity are influenced by discourse recipients' knowledge of speech act sequencing. Let's consider the impact of another type of pragmatic discourse knowledge in the following discourse that occurred during my second week on the faculty at the Air Force Institute of Technology:

Example 2.8.

Col./Dean: How are you settling in?

Campbell: Well ... I really think the students are exceptional. I'm excited
                 about the possibility of working on ...

Col./Dean: We want you to know how happy we are to have you on board.
                 [getting up] Thanks for coming by.

As the discourse recipient, I first perceived the Col./Dean's last sentence as incoherent because it did not appear to continue the discourse in which I thought we were engaged. He had requested an appointment with me and, after his first contribution, I assumed this interchange fell under the text type "chat." However, when he rose and thanked me for coming without contributing anything else to the discourse, I could no longer see the text type as chat because I know that this type of discourse requires participants to share equally in the exchange. As a very recent initiate in the military subculture, the discourse remained incoherent to me until a few days later when describing the experience to some of my students who are military officers. They recognized the text type of my discourse with the Col./Dean immediately, calling it the "welcome aboard" briefing. I have now added a new text type to my repertoire of pragmatic discourse knowledge.

Much research on reading comprehension notes that some of the knowledge recipients bring to the task of discourse processing is organized in terms of schema (e.g., Anderson & Pearson, 1984) or frames, scripts, plans, and goals (e.g., de Beaugrande & Dressler, 1981). Within these theoretical models, the incoherence of my conversation with the Col./Dean was a result of my inability to place this experience within my schema for text types or my lack of a frame for the text type "military welcome aboard." Thus I perceived a lack of continuity (hence coherence) in my discourse with him until I had the knowledge necessary to add this text type to my schema or to build the relevant frame.

Socially Constructed (Non-discourse) Knowledge

I noted ealier that socially constructed knowledge of discourse (e.g., speech acts and text types) may be involved in a recipient's sense of discourse continuity. Of course, other types of socially constructed knowledge can also be involved. For example, consider the following discourse:

Example 2.9.

A woman was murdered in this room last year. The police suspected her husband. The butcher knife turned up in the garden two months after her death. But the real killer was never found. (Cooper, 1982, p. 123)

Note that the propositions in this discourse are not explicitly related. In this case, however, the knowledge needed to construct a bridge that unifies this discourse is related not to knowledge of language or discourse, but to knowledge of the world. As Cooper (1982) explained:

[This is] an example of a text that invokes the maxim of relation in that its individual propositions do not seem to cohere. . . . Propositions that remain implicit in this text are that the woman's husband was suspected of her murder and that the murder was done with a butcher knife. Readers . . . assume the sentences are meant to be related and use the "murder" frame evoked in the first sentence to assign roles to the husband and the butcher knife and reasons for the action of the police. (p. 123)

As another example of the influence of knowledge not related to language or discourse, consider the Gantt chart in Fig. 2.5. Note that the activities for the fiscal year (FY) 1992 are given before 1991. In this case, the lack of discourse continuity is a result of knowledge of the Arabic number system used in our culture (and also of our left-to-right convention).

Interestingly, Grice (1975) made it clear that his maxims related to discourse "have their analogues in the sphere of transactions that are not talk exchanges . . ." (p. 47). He noted, "one of my avowed aims is to see talking as a special case or variation of purposive, indeed rational, behavior . . ." (Grice, 1975, p. 47). Grice (1975) used the following example to illustrate, "if I am mixing ingredients for a cake, I do not expect to be handed a good book, or even an oven cloth (though this might be an appropriate contribution at a later stage" (p. 47). Thus, the application of Grice's maxim of relation to knowledge not specifically related to discourse or language seems well motivated.

Idiosyncratic Knowledge

Up until now, I have considered the types of knowledge that, by definition, are shared by at least some subset of discourse participants. Of course, some recipient knowledge is not shared by others and is, therefore, idiosyncratic and unpredictable. One of my earlier examples illustrates the potential effect of such knowledge on our sense of discourse continuity.

Example 2.10.

The haystack was important. The cloth had ripped. (adapted from Clark & Haviland, 1977, p. 33)

I noted earlier that the only way a recipient might perceive this short discourse as unified is if he or she could build a bridge using the unstated topic "parachuting." Thus, for my readers, this discourse was incoherent until I noted that topic. In contrast, for me, this discourse was unified as I typed it in Example 2.10 because I already knew the topic. The point here is that, as a recipient, my idiosyncratic knowledge of the topic made the discourse coherent.

To summarize this section on relation and continuity, I have shown that the Gestalt principle of continuity is analogous to Grice's concept of relation. These concepts help us explain a recipient's sense of the continuity and coherence of discourse. I have also noted that discourse relation/continuity can be explicit or implicit, and that when it is implicit, the recipient who finds the discourse unified has built a bridge that provides a logical connection between successive discourse

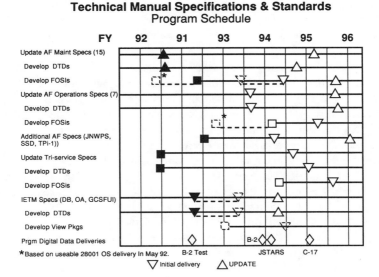

FIG. 2.5.   A Gantt chart illustrating the effect of socially constructed knowledge on a recipient's sense of discourse continuity.

contributions. In addition, I have recognized that discourse continuity is gradable and variable. And, finally, I have noted that a recipient's sense of discourse continuity is influenced by his or her knowledge. This includes semantic knowledge (i.e., word and sentence meaning), non-semantic, linguistic knowledge (i.e., phonology, morphology, and syntax), pragmatic discourse knowledge (e.g., speech act ordering and text types), non-discourse, socially constructed knowledge (e.g., frames), and idiosyncratic knowledge (i.e., unsystematic and unpredictable).

CONTINUITY VERSUS OTHER PRINCIPLES OF COHERENCE

In this section, I outline the three additional principles of discourse coherence that are suggested by Grice's work, but that appear to have no analogues in Gestalt theory: manner, quantity, and quality. Thus, I claim that discourse continuity is but one aspect of discourse coherence.

Manner

Grice proposed that discourse recipients expect other participants to make contributions that are orderly, unambiguous, and not obscure. As an example, consider the following short discourse:

Example 2.11.
John and Bill entered the room. Suddenly, he ran over to the plate on the floor
and licked up all the dog food on it. (Clark & Haviland, 1977, p. 18)

Note that lack of coherence in this brief discourse involves the lack of a unique
antecedent for the pronoun *he* in the second sentence: it could refer to either
John or *Bill*. (And, of course, it might also refer to an unnamed person, in which
case the discourse is still incoherent.) In other words, because the recipient cannot
identify the referent, the contribution made by the second sentence is seen as
unclear and, therefore, relatively incoherent.

Interestingly, the problem is resolved quite easily by bridging if the discourse
recipient knows that Bill is John's dog. In this case, it is a kind of idiosyncratic
world knowledge that permits bridging in order to create a clear and coherent
discourse. In short, a discourse recipient who can identify all text elements, (e.g.,
referents of pronouns), will perceive a discourse as more coherent than a discourse
recipient who cannot. Of course, other researchers have noted the importance of
clarity in describing coherence (e.g., Ziff [1984, p. 45] noted the importance of
"identifiability").

The coherence principle of manner appears to be distinct from the phenomenon
of discourse continuity. In other words, I would characterize the problem with
Example 2.11 as one of clarity but not as one of continuity or unity. In the case
of the principle of manner, incoherence appears to be the result of two or more
incompatible unities, not of a lack of unity. Of course, as the Gestalt drawing in
Fig. 2.6 demonstrates, Gestalt theorists were very interested in ambiguity. Note
that this drawing must be considered ambiguous because it prompts two, distinct
perceptions: (a) a pie plate with two pieces remaining or (b) a view of two KKK
members from below. Despite this interest in visual ambiguity, however, Gestalt
theory does not appear to provide a concept analogous to that of Grice's manner.
This point can be clarified by consideration of another example.

Consider the following discourse from a scientific article:

Example 2.12.
The /ŋg/ sequences have undergone three changes:
(1) /g/ → /k/ / /ŋ/_____ #
(2) /k/ → ɸ / /ŋ/_____ # in SG, wherever /k/ < /g/
(3) /g/ → ɸ / /ŋ/_____ /ə/. (Parker, 1980, p. 260)

No doubt some of my readers find this discourse excerpt incoherent. Grice's
maxim of manner suggests one explanation for this perception. Specifically, the
discourse recipient who cannot identify the graphemic or typographic symbols
(e.g., the "meaning" of ŋ) judges the discourse unclear and incoherent. In contrast,
those recipients with knowledge of phonological theory can identify the
graphemic symbol ŋ as the velar nasal phoneme occurring as the last segment

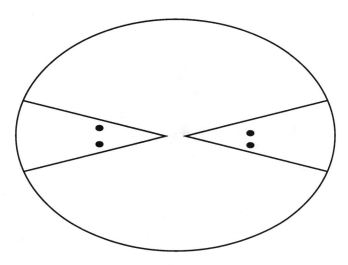

FIG. 2.6.   A Gestalt figure demonstrating ambiguity in visual perception. From "Literary vs. Technical Writing: Substitutes vs. Standards for Reality," by A. D. Manning, 1988, *The Journal of Technical Writing and Communication, 18*(3), p. 248. Copyright © 1988 by the Baywood Publishing Company, Inc. Reprinted by permission.

in the word *sing.* Thus, it is the discourse recipient's ability to identify a graphemic symbol that predicts his or her perception of coherence in this example.

A very similar kind of knowledge is involved in recipients' perceptions of clarity in the discourse reproduced in Fig. 2.7. No doubt most of my readers sense little clarity in this example. In this case, it is a recipient's socially constructed knowledge about graphemic symbols that determines clarity. As Benson and Burnett (1992) noted, "[a] native reader of Chinese . . . could not only easily recognize word units, but could also interpret the meanings of the punctuation, including the open circle for 'period,' [and] two different kinds of 'commas' . . ." (p. 94). Therefore, the recipient's knowledge of the social conventions of relatively small groups (e.g., of linguists) and of much larger groups (e.g., of literate Chinese) determine his or her sense of discourse clarity.

Inasmuch as clarity and continuity are aspects of discourse coherence, it is not surprising that they share many characteristics. For instance, our sense of discourse clarity depends on a range of knowledge (e.g., graphemic characters, semantics, etc.). In addition, our sense of discourse clarity is variable depending on the particular knowledge of the recipient (e.g., those who know the graphemic characters vs. those who do not). Furthermore, our sense of discourse clarity is gradable: the more graphemic symbols identified, the stronger a recipient's sense of clarity and coherence.

Despite these similarities, note that we cannot claim that Example 2.12 is ambiguous. In other words, it seems impossible to argue that this example

魔鬼的想象

第一次见着她的时侯，他便觉着这是个挺浪漫的小姑娘。

有意无意间，他挑些不一般的经历告诉她：

……那几年，才五、六岁，就跟着父母到处流浪……到了刮风下雨的夜晚，只好钻进载煤的货车皮里过夜……唉——那些日子可苦了！

她想想，也真的很苦。便很钦佩地望望他。

后来，熟悉一些了的时侯，他说着说着就不知怎么的漏了嘴：

其实，那些日子过起来倒并不见得有听起来可怕，想当初年纪又小，就那么稀里糊涂地过来了……现在反倒成了笔可供吹嘘的资本了——也算是个"吃过苦"的人了！

她忍不住想笑。

过后她又生他气。

可他不以为然：

——你呀，小说读得太多，习惯于接受夸大了的人类情感……你去看看生活中有多少人，度过了远比小说要更真实地痛苦和残酷的时刻，也就——不过如此。

FIG. 2.7. A discourse example illustrating the influence of socially constructed knowledge on a recipient's sense of coherence related to manner or clarity.

suggests a variety of incompatible unities. Thus, the recognition of ambiguity in Gestalt theory appears unrelated to examples such as 2.12 and, therefore, I cannot argue that Gestalt theory provides a concept analogous to that of manner, only that the theory recognizes the result of some violations of manner. The bottom line here is that the concept of manner is not directly related to discourse continuity and therefore is not central to the theme of the remaining chapters in this book.

Quantity

Grice proposed that recipients expect that other discourse participants will contribute the appropriate amount of information, neither too much nor too little for the purposes of the discourse. As an introductory example, consider the following brief discourse:

Example 2.13.
Kenny: What are you reading?
Tom: A book. (adapted from Parker & Riley, 1994, p. 13)

Note that this discourse appears to lack coherence inasmuch as Tom's contribution is perceived as incomplete.

In addition, I should note that, although this discourse lacks explicit completeness, a recipient might easily perceive implicit completeness by bridging. Parker and Riley (1994) provided a likely line of reasoning for Kenny to construct a bridge:

- I asked Tom what he was reading and the context of my question required him to tell me either the title of the book or at least its subject matter. . . .
- [Tom] told me what I could already see for myself . . . [he is] giving me less information that the situation requires. . . .
- [Tom] does not want to be disturbed (and thus is trying to terminate the conversation). (p. 13)

Thus, Kenny constructs a bridge that accounts for the incompleteness of Tom's contribution, thereby creating a coherent discourse. For such a recipient, the discourse is seen as coherent despite the superficial incompleteness. The fact that this bridging seems automatic in Example 2.13 does not render it any less necessary. It bears repeating that the harder recipients must work to construct a bridge, the less coherence they sense in the discourse.

In the previous case, it is the discourse recipient's knowledge of pragmatics (i.e., that *a book* is incomplete or at least inadequate as a response in this interchange) that is involved in the perception of coherence. However, we find that the same range of knowledge types may be involved in a recipient's sense of coherence involving either continuity/relation or quantity. As one example of the influence of a recipient's linguistic knowledge on coherence involving adequacy, consider the following two versions of a common children's rhyme:

Example 2.14a.
Mary and Johnny sittin in a tree,
K-I-S-S-I-N-G.
First comes love; then comes marriage;
Then comes Johnny with the baby carriage.

Example 2.14b.
Mary and Johnny sittin in a tree,
K-I-S-S-I-N-G.
First comes love; then comes marriage;
Then comes Johnny with the baby.

Note that Example 2.14a is clearly more coherent than Example 2.14b. The concept of quantity explains that difference by focusing on the incomplete meter and rhyme (i.e., phonological elements) of the last line in Example 2.14b. Thus, linguistic knowledge of phonology influences a recipient's sense of coherence involving the quantity of information in the discourse.

As an example of how a recipient's discourse knowledge can affect his or her sense of coherence involving quantity, consider two sets of sample headings from a scientific article:

Example 2.15a.
ABSTRACT
INTRODUCTION
MATERIALS AND METHODS
RESULTS
DISCUSSION
REFERENCES

Example 2.15b.
ABSTRACT
INTRODUCTION
RESULTS
DISCUSSION
REFERENCES (adapted from Paccamonti, Chang, Drost, Wilcox, Prichard, & Fields, 1990)

Note that the set of headings in Example 2.15a demonstrates all of the typical sections of an experimental report and (with the appropriate content) suggest a coherent discourse (in that respect). In contrast, the set of headings in Example 2.15b suggests a partial discourse because no "materials and methods" heading is given. The point here is that the concept of quantity is one explanation for the different senses of coherence between these examples. Those of us who sense this difference are influenced by our knowledge of the text type "experimental research article."

Quality

Grice proposed that discourse recipients expect other participants not to produce discourse that is false or for which the participants' lack adequate evidence. As an example, consider the following discourse, which is overheard by a third party:

Example 2.16.

(Tom & Frank's) Boss:	What's Frank's home phone number?
Tom:	1 - 2 - 3 - 4 - 5 - 6 - 7

Note that this discourse is perceived as incoherent if the third party recipient knows that (a) Tom has given his boss the wrong phone number, and (b) Tom knows the correct phone number. In other words, it is a discourse recipient's idiosyncratic (i.e., unsystematic, hence unpredictable) knowledge of "world facts" that determines the coherence of this discourse. On the other hand, if the third party recipient does not know that Tom has given his boss the wrong phone number, then the recipient interprets the discourse as accurate and coherent.

Inasmuch as accuracy is another coherence principle, it shares a number of characteristics with the other principles I have discussed. For instance, our sense of accuracy may be implicit rather than explicit; in other words, a recipient may construct a bridge in order to make the discourse coherent despite its explicit inaccuracy (e.g., the recipient may use his or her knowledge of Tom, Frank, and their boss to judge that Tom has purposely given inaccurate information because Frank has asked Tom not to give their boss his number). In addition, our sense of accuracy is obviously gradable or relative. Moreover, our sense of discourse accuracy depends on many types of knowledge (e.g., semantics, idiosyncratic world knowledge, etc.) and will be variable depending on the particular knowledge of a recipient (e.g., whether he or she knows Frank's number).

As another example, consider the following discourses:

Example 2.17a.
Kathy: I'm gonna kiss you.
Frank: Thanks.

Example 2.17b.
Kathy: I'm gonna hit you.
Frank: Thanks.

Note that Example 2.17a appears to be coherent, whereas Example 2.17b does not. Specifically, Frank's contribution is seen as an accurate contribution in Example 2.17a but not in Example 2.17b. Specifically, Kathy's contributions

TABLE 2.1

Felicity Conditions for Performing Each of the Six Categories of Speech Acts

	Preparatory	Sincerity	Essential	Propositional Content
Representative: assertion	1. S believes H doesn't know P.	1. S believes P.	1. Counts as an assertion of P.	1. Any P.
Directive: request	1. S believes H able to do A. 2. A is something H would not normally do.	1. S wants H to do A.	1. Counts as attempt to get H to do A.	1. Future A of H.
Question: question	1. S does not know P. 2. P is something H would not normally provide.	1. S wants to know P.	1. Counts as attempt to elicit p from H.	1. Any p.
Commissive: promise	1. S believes H wants A done. 2. A is something S would not normally do.	1. S intends to do A.	1. Counts as obligation to do A.	1. Future A of S.
Expressive: thanking	1. S believes A benefits S.	1. S feels appreciation for A.	1. Counts as expression of appreciation for A.	1. Past A of H.
Declaration: naming	1. S has the authority to name X	1. S intends to name X.	1. Counts as naming of X.	1. Name for X.

Note. From *Linguistics for Non-Linguists: A Primer With Exercises, Second Edition* (pg. 17) by F. Parker and A. Riley, 1994, Boston, MA: Allyn & Bacon. Copyright © 1994 by Allyn & Bacon. Reprinted by permission.

appear to signal a promise in Example 2.17a and a threat in Example 2.17b. However, Frank's contribution of *thanks* in both examples signals that he takes the speech act performed by both of Kathy's contributions as that of a promise.

Thus, in the case of Example 2.17b, the lack of coherence and accuracy is dependent on the recipient's pragmatic discourse knowledge: what counts as a well-formed (technically known as "felicitous") speech act. To explore this further, Table 2.1 lists the conditions for one of each of the six categories of speech acts: representatives, directives, questions, commissives, expressives, and declarations. Using the conditions for "promising" under the category *commissives*, note that when we believe Kathy's contribution in Example 2.17a constitutes a promise, we assume that all of the following conditions are met: (a) Kathy believes that Frank would like to be kissed by her; (b) kissing Frank is something Kathy does not ordinarily do; (c) Kathy actually intends to kiss Frank; and (d) this kissing has not yet happened and Kathy will be the agent of this kissing. If any of these conditions is unmet, then we do not believe the utterance in question is a promise. Thus, because Kathy's contribution in Example 2.17a meets all of the conditions just given, we believe she is making a promise and that Frank's contribution of *thanks* is an accurate response.

In contrast, consider Example 2.17b again. Note that, when we believe Kathy is making a threat, we believe that all the following conditions are met: (a) Kathy believes that Frank would **not** like to be hit by her; (b) hitting Frank is something

Kathy does not ordinarily do; (c) Kathy actually intends to hit Frank; and (d) this hitting has not yet happened and Kathy will be the agent of this hitting. Thus, the small (but obviously significant) change in condition (a) is the only difference between a promise and a threat.

In Example 2.17b, Frank's contribution is not seen as accurate unless the discourse recipient can construct a bridge that explains the apparent inaccuracy of his response. (This is what we are doing when we assume that Frank's response is sarcastic, and that he actually means the opposite of what he literally states.) The point here is that discourse recipients use their knowledge of conditions on speech acts to judge the coherence of discourse contributions involving accuracy.

As with coherence involving manner and quantity, coherence involving the quality of information appears to have no analogous principle in Gestalt theory. Again, I doubt that many readers would characterize the problem in discourse Examples 2.16 and 2.17b as one of continuity or unity. In order to make this point more clearly, it may be instructive to consider the different natures of the phenomena I have been comparing in this chapter—visual/auditory phenomena on one hand and discourse on the other. In particular, we must consider the fact that, although discourse is necessarily semiotic (i.e., discourse is, by definition, symbolic or meaningful), visual and auditory objects may or may not be. As an example, consider the two visual objects depicted in Fig. 2.8. Note here that (a) is perceived as semiotic, whereas (b) is not. Interestingly, work by textlinguists suggests that when we perceive a visual object as coherent (i.e., semiotic), it becomes a kind of text (e.g., Dorfmüller-Karpusa & Dorfmüller, 1985). The point here is that Gestalt theory was developed to explain the nature of our perception of non-semiotic visual and auditory phenomena and, therefore, cannot provide a comprehensive theory of discourse coherence. Nevertheless, it is the premise of this book that Gestalt theory can provide a comprehensive theory of one aspect of discourse coherence–continuity. I explore this issue further in the following chapter.

SUMMARY

This chapter relied on Grice's cooperative principle and maxims in order to develop a theory of coherence because his work outlines discourse participants' expectations, which are central to their sense of coherence in a discourse. I explored a categorization of recipients' expectations consisting of four coherence principles: relation, manner, quantity, and quality. In addition, I showed that all explicitly incoherent discourse can be construed as implicitly coherent by a recipient with the ability to build a bridge explaining away the explicit "flaw" in relation, quantity, manner, or quality. Furthermore, I noted that a recipient's sense of coherence is gradable (i.e., recipients sense relative degrees rather than its absolute presence or absence) and variable (i.e., each recipient has a unique set of knowledge, experience, and motivation that affects his or her sense of coherence). Moreover, I

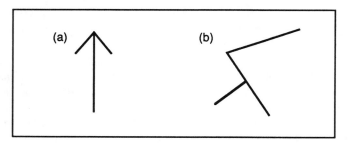

FIG. 2.8.   A semiotic visual object (a) and a non-semiotic (non-symbolic) visual object (b) composed of three lines.

considered the full range of recipient knowledge that may influence his or her sense of coherence: semantic knowledge, other linguistic (non-semantic) knowledge, pragmatic discourse knowledge, other socially constructed or culturally transmitted (non-discourse) knowledge, and idiosyncratic knowledge.

Finally, I used Gestalt theory to argue that a recipient's sense of coherence where he or she sees a relation between discourse contributions is analogous to the innate, human sense of unity where there is visual continuity. In contrast, I claimed that the other coherence principles—manner, quantity, and quality—are not involved in a recipient's sense of discourse continuity because they appear to have no analogues in Gestalt theory. The scope of the rest of this book is, for the most part, limited to one aspect of coherence—continuity.

Clearly, I ignored the obvious truth that recipients are routinely influenced by all of the previously discussed coherence principles at once in a discourse. Although my approach has been taken primarily for the sake of clarity in exposition, this fact should not bother us because, in principle, all of those influences could be described by the principles set forth in this chapter. In fact, I turn to a discussion of the interaction of principles of coherence and cohesion in chapter 4. But first I must explore the applicability of Gestalt principles to the phenomenon of cohesion in chapter 3.

CHAPTER 3

Continuity and Local Cohesion:
The Producer's Perspective

In the previous chapter, I outlined four coherence principles that predict when a recipient will sense coherence in discourse, as well as the types of knowledge that affect his or her sense of coherence. I demonstrated that one of these principles has a direct correlate in Gestalt theories of unity in visual and auditory phenomena, and concluded that Grice's concept of relation describes a recipient's sense of one aspect of discourse coherence—continuity. In this chapter, I consider only that one aspect of coherence. Thus, I argue that cohesion and coherence involving continuity are closely related.

To begin, I discuss the shift in perspective in this chapter from discourse recipients to discourse producers, which entails a shift from discourse coherence to cohesion. In the remainder of this chapter, I consider those cohesion principles that explain the production of a sense of coherence based on continuity; these include the principles of similarity and proximity.

RECIPIENTS VERSUS PRODUCERS

In chapter 2, we considered coherence from the discourse recipient's perspective and discovered that it is variable depending on the motivation and knowledge of each recipient. Defining coherence in terms of recipient response is quite common. For instance, I noted in the previous chapter that a number of researchers have defined coherence in terms of recipient expectations (e.g., Fahnestock, 1983; Witte & Faigley, 1981). Interestingly, this definition of coherence is implicit in some previous experimental studies that categorize the written discourses of college freshmen as coherent/incoherent based on the holistic (i.e., subjective) judgments of a few rhetoric teachers (e.g., Tierney & Mosenthal, 1983).

Of course, some research concludes that, because recipient responses can and do vary according to the motivations and knowledge of recipients, theories of cohesion (which are by definition concerned with cues placed in a discourse by a producer) are superfluous as indicators of coherence (e.g., Bamberg, 1983; Tierney & Mosenthal, 1983). All of this research has, to my knowledge, been based on the theory of cohesion set out in Halliday and Hasan's (1976) seminal work, *Cohesion in English.* In the years following publication of that work, it became increasingly suspect for writing specialists to focus on the discourse itself (see Parker & Campbell, 1993, for a detailed discussion of the misunderstanding between writing specialists and linguists that has encouraged this situation). This point is especially well made by remembering that, for a number of years, product-based research (and to some extent pedagogy) was almost completely neglected in favor of process-based research and pedagogy among writing specialists.

By the late 1980s, such neglect was clearly viewed by many as a mistake. For instance, Kinneavy (1987) wrote that "the process revolution . . . has not been implemented without some problems. One of these has been the neglect and disregard . . . of almost any concern with product at all" (p. 1). Perhaps even more descriptive of this change in attitude is the title of a forum piece in 1988 in the *Technical Writing Teacher*: "I Was a Victim of the Process Approach" (Dukes, 1988).

It seems crucial to me that technical communication specialists keep in mind that technical or scientific discourse (which are both clearly communicative rather than, for example, expressive) require both a receiver and a producer, with the discourse itself as the only observable connection between these participants. In addition, it is clear that recipients do not have complete freedom in interpreting discourse. As de Beaugrande and Dressler argued in 1981:

> We must guard against allowing the text to vanish away behind mental processes. Recent debates over the role of the reader point up the dangers of assuming that text receivers can do whatever they like with a presentation. If that notion was accurate, textual communication would be quite unreliable, perhaps even solipsistic. There must be definitive, though not absolute, controls on the variations among modes of utilising a text by different receivers. (p. 35)

Clearly, then, discourse continuity and coherence are **both** (a) perceptions of a human recipient confronting a discourse and (b) qualities of that discourse. It is natural for recipients of discourses to state that a particular set of instructions "isn't unified" or that a particular section of an experimental report "really hangs together" despite the fact that other recipients may not perceive these qualities in the same instructions or report.

The remainder of this chapter focuses on those indications or cues placed in a discourse by its producer that appear to influence a recipient's sense of continuity.

COHESION: ESTABLISHING CONTINUITY

In chapter 2, I noted that continuity is one principle explaining recipients' sense of discourse unity and coherence. In this section, I want to take up the question of **how** that sense of continuity is influenced by a producer. Consider Fig. 3.1 again, which appeared in chapter 2. One of the things we noted about this figure was that line XD is naturally perceived as the continuation of line AX, whereas line XC is not. Thus, AXD is perceived as unified, whereas AXC is not. The following discussion outlines a sub-principle of continuity that explains how this continuity is established.

The Cohesion Principle of Similarity

Note that, in Fig. 3.1, segments AX and XD demonstrate similar form (i.e., waviness), whereas segment XC demonstrates dissimilar form (i.e., straightness). These similarities are one perceptual cue determining which segments are acceptable continuations, thereby creating a sense of visual cohesiveness, which results in the perception of continuity and unity.

The principle of similarity applies to auditory as well as visual phenomena. To illustrate, consider the musical examples in Fig. 3.2, which were suggested by the work of Lerdahl and Jackendoff (1983). Note that our most natural perception of the sounds represented by Fig. 3.2a is as one unit. In contrast, our auditory interpretation of Fig. 3.2b is as two units: a group of two notes followed by a group of three. The principle of similarity explains this perception by acknowledging the impact of similar pitch as a cohesive cue that influences our

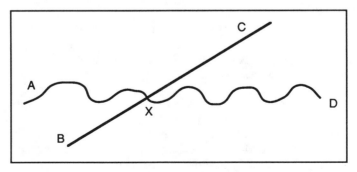

FIG. 3.1.   A Gestalt figure demonstrating that the principle of similarity is the foundation for the principle of continuity.

(a)

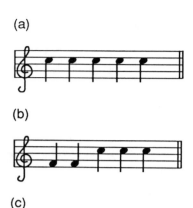

(b)

(c)

FIG. 3.2. A musical example demonstrating that the principle of similarity is the foundation for the principle of continuity in auditory phenomena.

perceptions of auditory continuity and unity. In addition, Fig. 3.2c illustrates that
the cohesion principle of similarity, like the coherence principles discussed in
chapter 2, is gradable. In other words, we perceive the unity of the first two
notes more strongly in Fig. 3.2b than in 3.2c because of the greater degree of
similarity in 3.2b compared to 3.2c.

To illustrate how the principle of similarity explains some of a recipient's
sense of continuity in discourse, consider the following example, which appeared
in chapter 1.

Example 3.1.
Mr. Krishan Saigal, P.E., will serve as Lead Engineer. Mr. Saigal's primary tasks
will include:

• Plan and provide direction for technical work elements.

• Coordinate technical direction of subcontractors.

• Assist in coordinating and disseminating project-related information to the Project
Team . . .

Mr. Saigal will also serve as Construction Manager for the Project Team, with the
following primary responsibilities . . . (SCS Engineers, 1991, p. 2-2)

Note that the bulleted portion of Example 3.1 produces a strong sense
of continuity. Note also that the producer has used a number of similar discourse
elements within this segment of the example: for instance, similar propositional
content (i.e., actions typical of a managing engineer), similar sentence structure
(i.e., syntactic parallelism), and similar visual cues (i.e., use of "bullets" and
same page placement). The point here is that the use of these similarities is
cohesive in that they produce a sense of continuity in this discourse excerpt.

The Range of Cohesive Discourse Elements

In the discourse example just given, I noted the impact of a range of discourse elements: from semantic to syntactic to visual. One of the justifications for the theory of cohesion presented in this chapter is precisely this range of elements that producers manipulate in order to affect coherence. Halliday and Hasan's theory of cohesion, the earliest and most comprehensive to date, considers only the impact of semantic discourse elements. The authors stated that they were "excluding from consideration the effects of formal devices such as syntactic parallelism, metre and rhyme . . ." (Halliday & Hasan, 1976, p. 10). Some research has attempted to expand Halliday and Hasan's theory (e.g., Hartnett, 1986; Markels, 1983; Stotsky, 1983). As I have argued elsewhere,

> Despite the need for this kind of research, Markel's paper proposes only one new category of cohesion (syntactic). And both Stotsky and Hartnett modify Halliday and Hasan's original categories. In essence, no research has established the range of . . . cohesive devices available to writers. (Campbell, 1991, p. 223)

In contrast, the theory of cohesion developed in this chapter explains the impact of all discourse elements. To this end, the following discussion catalogues the four categories of discourse elements that producers manipulate. Despite the differences in types of elements that are involved in creating discourse continuity (e.g., only the last category deals with semantic elements, whereas the first three deal with varieties of non-semantic elements, including non-linguistic elements), I intend to show that the principle of similarity underlies the unifying characteristics of them all.

Visual Elements

Research in educational psychology and in technical communication has long recognized the importance of visual cues for producing coherent discourse. Although perhaps less recognized, visual cues are used not only in textbooks and technical documents, but in a wide variety of text types. For instance, consider the following excerpt from Tolkien's (1966) classic novel, *The Hobbit*.

Example 3.2.
But suddenly Gollum remembered thieving from nests long ago, and sitting under the river bank teaching his grandmother, teaching his grandmother to suck— "Eggses!" he hissed. "Eggses it is!" Then he asked:

> *Alive without breath,*
> *As cold as death;*
> *Never thirsty, ever drinking,*
> *All in mail never clinking.* (p. 76)

Note that the last four lines produce a strong sense of continuity and thus comprise an obvious unit within the discourse. The similar visual characteristics that prompt

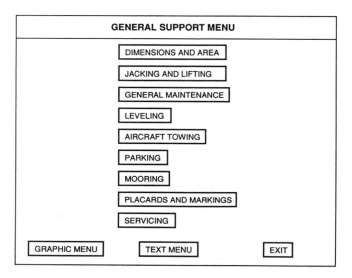

FIG. 3.3.   A computer screen illustrating the role of visual similarities in creating a sense of discourse continuity in hypertext communication.

this perception include the similar page placement (i.e., indention) and italic print of all four lines. (Obviously, other similarities are also involved, but I will delay our discussion of the interaction of elements until chapter 4.)[1]

As another example of discourse continuity that is produced by visual similarities, consider Fig. 3.3, a screen from the hypertext version of the United States Air Force technical orders for the F-15E aircraft. Consider the 12 entries (i.e., "DIMENSIONS AND AREA" through "EXIT") that appear on the screen. Note that for all 12 the visual display includes a surrounding rectangular box, and (in the original) white letters on blue background. Thus, the principle of similarity provides one explanation for the cohesiveness that produces a sense of continuity in these discourse segments. Moreover, note that for nine of the entries screen placement is also similar; all appear with a flush left margin roughly in the center of the screen. Thus, the principle of similarity provides an explanation for the greater cohesiveness of these nine entries (and also the sense of the discontinuity between them and the three other entries because, for example, *TEXT MENU* is roughly centered in the screen but not flush with the left margin of the nine entries above it).

I also explored (Campbell, 1992) the production of cohesion in technical graphics types: charts, graphs, diagrams, and tables. Not surprisingly, part of the cohesiveness in graphics is the result of similar visual elements. For example, consider Fig. 3.4. Note that the producer of this graphic used a variety of visual

[1]For those whose interest is piqued and who do not have their copy of *The Hobbit* nearby, the answer to Gollum's riddle is "fish."

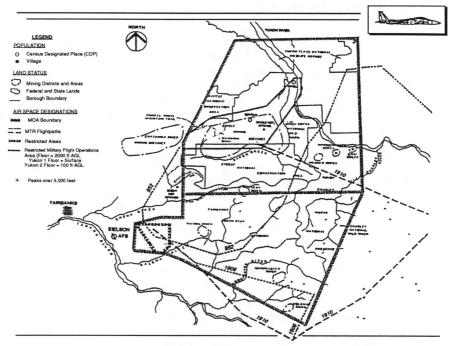

FIG. 3.4. A map illustrating the role of visual similarities in creating a sense of discourse continuity in written, technical graphics.

characteristics for the lines in this map. For instance, according to the legend, all "Land Status" designations are produced with variations of solid lines, whereas all "Airspace Designations" are produced with variations of broken lines. The point here is that it is the repetition of similar visual form that produces cohesiveness and a sense of continuity in this map.[2]

In sum, visual elements that producers manipulate in order to create a sense of continuity include similar page or screen placement (e.g., indention), typography (e.g., italics), color (e.g., white letters), and visual form (e.g., geometric shapes and shape of lines). I am not claiming here that visual elements alone create a recipient's sense of continuity in discourse. I want to assure my readers that I am treating each type of cohesive element separately in this chapter for clarity in exposition. I discuss the interaction of these cohesive elements and the coherence principles in chapter 4.

---

[2]See Campbell, 1992, for a more thorough description of cohesion in technical graphics. See Bernhardt, 1986, or Moore and Fitz, 1993, for a description of Gestalt principles that provide a cogent theoretical basis for understanding the visual aspect of document design.

Auditory/Phonological Elements

Now let's turn to auditory/phonological discourse elements. Consider Fig. 3.3 again. Imagine that, each time a recipient "clicks on" one of the nine entries discussed earlier, the computer produces two short beeps, whereas clicking on one of the three other entries along the bottom of the screen results in one long beep. In this case, the cohesion principle of similarity explains the effect of similar auditory cues used by the producer to create a sense of continuity.

The principle of similarity is also the foundation of well recognized unifying devices such as rhyme, meter, and alliteration. For example, the principle of similarity explains part of our sense of continuity in limericks such as the following one attributed to Edward Lear in 1846:

Example 3.3.
There was an Old Man who supposed
That the street door was partially closed;
But some very large rats ate his coats and his hats,
While that futile old gentleman dozed. (Abrams, 1986, p. 1591)

Note that the rhyming of lines one, two, and four is produced through the use of final words in which the last syllable is similar. Specifically, despite their various spellings, the coda of each word is pronounced /ozd/. In addition, these lines follow the same metrical pattern; in other words their rhythm and intonation is similar. Thus, the principle of similarity explains the continuity of this discourse by acknowledging the cohesiveness of its phonological elements.

Alliteration constitutes a different kind of phonological similarity as found in lines 3334–3338 from "The Miller's Tale":

Example 3.4.
In al the toun nas brewhous ne taverne
That he ne visited with his solas,
Ther any gaylard tappestere was.
But sooth to seyn, he was somdeel squaymous
Of fartyng and of speche daungerous. (Chaucer, 1390/1987, p. 70; emphasis added)

Note that the producer uses similar phonological structure in four of the words in the boldfaced line. Specifically, the initial sounds (i.e., onsets) of these words are similar. Thus, the principle of similarity also accounts for the cohesive influence of alliteration on a recipient's sense of discourse continuity.

Tannen quoted excerpts from Martin Luther King, Jr.'s "I Have a Dream" speech to demonstrate that poetic devices such as alliteration are used in discourse not necessarily classified as literary:

Example 3.5.
I have a dream
that my four little children
will one day live in a nation
where they will not be judged
by the color of their skin
but **by the content of their character.** (Tannen, 1989, pp. 83, 89; emphasis added)

Note that the phonological structure of the two boldfaced lines is similar. Specifically, the producer uses alliteration (repetition of similar onsets) in the words, *color*, *content*, and *character* to create a cohesiveness that prompts recipients to see the last line as a continuation of the previous one.

Yet another type of phonological similarity is involved in producing continuity in the following oral discourse excerpt from an Air Force briefing:

Example 3.6.
there's four procedures you always go through, uh before you service an aircraft.

[**first** is you determine whether the aircraft is either normal weight or heavyweight . . . **then** uh you determine what the ambient air temperature is . . . and **third** is uh you determine whether it's gonna need nitrogen or hydraulic servicing . . . and then the **fourth** thing is you, uh and as I said before you determine whether the aircraft is on jacks or not].

um basically uh that's it. (Swillum, 1992)

In the original oral version of this discourse, each of the boldfaced items was given a rising pitch and louder stress than the surrounding items. Thus, the cohesion principle of similarity describes the cohesiveness of phonological elements that the producer uses to create a sense of continuity in the middle segment of the excerpt just given.

In sum, auditory and phonological elements that producers manipulate in order to produce a sense of continuity include: non-linguistic elements (e.g., computer beeps), codas (i.e., rhyming), onsets (i.e., alliteration), rhythm, pitch, and stress (i.e., meter and prosodic emphasis).

Syntactic Elements

No doubt many readers noted that Dr. King also influenced a recipient's sense of continuity in Example 3.5 through his use of cohesive, similar syntactic structure (i.e., syntactic parallelism). Specifically, both of the boldfaced lines are composed of embedded prepositional phrases (i.e., *by x of y*). The following example from a scientific journal also demonstrates the cohesion property of similarity involving syntactic structure:

Example 3.7.
For five of the items, sentences exhibiting the greatest durational distinction
between morphemic and non-morphemic /s/ *were chosen.* **For the remaining three
items,** two sentences with no distinction, and one with reverse distinction . . . *were
selected.* (emphasis added, Walsh & Parker, 1983, p. 204; emphasis added)

Note that in this excerpt both sentences have an initial prepositional phrase
(highlighted in boldface) followed by a passive main clause (highlighted with
italics). The producers' use of repeated syntactic form creates cohesiveness
between the two sentences, thereby encouraging a sense of continuity in this
example.

Example 3.8 also demonstrates how a producer influences continuity through
the use of similar morphosyntactic elements (e.g., verb tense or aspect).

Example 3.8.
He does not even think **Surely Judith didn't write him about that letter** or **It
was Clytie who sent him word somehow that Charles has written her.** (Faulkner,
1936, p. 353; emphasis added)

Note that the past tense is used in the clauses in boldface (i.e., *didn't* and *was*),
whereas the present tense is used in the main clause (i.e., *does*). The cohesion
principle of similarity explains why there is a stronger sense of continuity between
the boldfaced clauses than between the first clause and the boldfaced clauses
that follow it. Thus, a recipient's sense of continuity is influenced by the use of
similar morphosyntactic elements.

At this point, it may be instructive to consider two versions of a scientific
discourse: one that demonstrates morphosyntactic cohesiveness and one that does
not:

Example 3.9a.
The dog **was placed** under general anesthesia and the left pelvic limb was prepared
for aseptic surgery. . . . [T]he tibiotarsal joint **was accessed** via osteotomy of the
medial malleolus. . . . Two fragments which comprised the proximal lateral ridge
were identified (figure [x]). (adapted from van Ee, Gibson, & Roberts, 1990, p.
2; emphasis added)

Example 3.9b.
The dog **is placed** under general anesthesia and the left pelvic limb was prepared
for aseptic surgery. . . . We **accessed** the tibiotarsal joint via osteotomy of the
medial malleolus. Two fragments which comprised the proximal lateral ridge **were
identified** (figure [x]).

Note that, in the original version Example 3.9a, the producers have used past
tense and passive voice in each of the main verbs in all three sentences, thus
influencing a recipient's sense of continuity in this discourse through the use of

similar morphosyntactic elements. In contrast, the altered version Example 3.9b produces a sense of less continuity precisely because dissimilar elements (i.e., present passive, past active, and past passive) are used.

In sum, syntactic and morphosyntactic elements that producers manipulate in order to produce a sense of continuity include: syntactic structure (i.e., parallel structure) and verb tense and aspect.

Semantic Elements

Halliday and Hasan's theory of cohesion posits four categories of semantic cohesion: reference, substitution and ellipsis, lexical cohesion, and conjunction. In this section, I argue that the principle of similarity can explain the cohesive effect of all of these categories except conjunction. Furthermore, I propose an explanation for why conjunction should be treated differently than the other types of cohesion.

Reference. The following example from a veterinary brochure illustrates cohesion produced through reference:

Example 3.10.
A soft rubber or soft plastic tube often becomes doubled in the esophagus or pharynx, and **it** may even come back through the mouth as **it** is being passed. This can be dangerous in that it may cause severe hemorrhage when withdrawn. (Adams, 1970, p. 2; emphasis added)

Note that the boldfaced items in Example 3.10 have a similar referent (i.e., *tube*). In other words, the same referent is repeated throughout this section of the discourse. As Halliday and Hasan (1976) noted, "[i]n the case of reference . . . the cohesion lies in the continuity of reference, whereby the same thing enters into the discourse a second time" (p. 31). Thus, the authors imply that repetition is central to the continuity of discourse.

The importance of reference (also called *anaphora*) in producing coherent discourse has been widely studied by those interested in natural language processing in artificial intelligence (e.g., Charniak, 1972; Chomsky, 1976; Grosz, 1977; Winograd, 1972). For instance, Sidner (1983) proposed a theory of discourse comprehension in which discourse "focus" is established and maintained partly through the appearance of referential expressions. In addition, Sidner (1983) recognized that referential expressions are a means of establishing relation or continuity in discourse,

Focusing and focus as they have been used here bear directly on Grice's concerns; for they suggest a means for carrying out the maxim of relevance. Namely, a speaker is speaking relevantly in a discourse if he or she introduces a focus, and proceeds to another one by mentioning it and re-mentioning it with definite anaphora. (p. 330)

Thus, because reference describes the repetition of similar referents, it appears that the cohesion principle of similarity introduced in this chapter can be used to explain the cohesiveness of reference, as well as its role in establishing discourse continuity.

Substitution and Ellipsis. The following examples illustrate cohesion produced through the use of substitution (3.11a) and ellipsis (3.11b):

Example 3.11a.
This is a fine **hall** you have here. I've never lectured in a finer **one**.

Example 3.11b.
This is a fine **hall** you have here. I've never lectured in a finer Ø. (Halliday & Hasan, 1976, p. 146; emphasis added)

As a native speaker of American English, I find Example 3.11b strange, but apparently this usage is unexceptional in British English. Note that ellipsis is also used ubiquitously and unexceptionally in typical American English responses such as *Eating* for *I am eating* after you've been asked, *What are you doing?* Note that the boldfaced items in both Examples 3.11a and 3.11b repeat the same referent (i.e., *hall*). In any case, in Example 3.11a, *one* substitutes for *hall*, whereas in Example 3.11b the null sign signifies the ellipsis and presupposition of *hall*. As Halliday and Hasan noted, substitution and ellipsis are, in many ways, like reference in that all three types of cohesion involve the continuation of a previously stated referent. Like reference, then, the cohesiveness of substitution and ellipsis appears to be covered by the cohesion principle of similarity introduced in this chapter.

Lexical Cohesion. As one example of this type of cohesion, consider the following dialogue between a doctor and his farmer-patient in a short story:

Example 3.12.
[Doctor:]"But the sure thing is you've got to cut out **farm work**. You can <u>feed the stock</u> and <u>do chores about the barn</u>, but you can't do **anything in the fields** that makes you short of breath."
[Farmer-Patient:] "How about *shelling corn*?"
[Doctor:]"Of course not!" (Cather, 1987, p. 1044; emphasis added)

Let's consider the set of rather complex, semantic relationships that hold between the phrases I have highlighted in Example 3.12. First, note that the boldfaced phrases are synonyms. Second, note that the underlined phrases exhibit overlap with each other (i.e., part of their meaning is the same), and that they are antonyms of the boldfaced phrases (i.e., the meaning of *farm work* is excluded from the meaning of *do chores about the barn*). Third, note that the exact semantic relationship between the italicized phrase and the other highlighted phrases is in dispute: its meaning might exhibit overlap with the underlined phrases (the farmer's preference) or it might be an antonym of the underlined phrases (the doctor's preference).

This rather detailed examination makes two points: (a) talking about semantic relationships is difficult, and (b) the cohesive effect of lexical items can be the result of their semantic relationships. The first point is clear to anyone who has taken a course or read a book on semantics, and is no doubt largely responsible for the fact that relatively little progress has been made in semantic theory compared to the other areas of linguistics. The second point further corroborates my claim that the cohesion principle of similarity explains the unifying effect of the full range of discourse elements, including semantic ones.

Lexical cohesion can involve other types of semantic relations. For example, consider the following list of sentences from various sections of a short story.

Example 3.13a.
One time there used to be a field **there** in which they used to play every evening with other people's children.

Example 3.13b.
Her father used often to hunt them **in out** of the field.

Example 3.13c.
The organ-player had been ordered to **go away** and given sixpence. (Joyce, 1946, pp. 46, 50; emphasis added)

Note that all of the boldfaced words are deictic terms (i.e., they signify spatial relationships between other semantic elements and the speaker). In this example, the boldfaced terms anchor all objects and events in the discourse in relation to a character's home. For instance, the demonstrative *there*, the verb *go*, and the preposition *away* mean "away from Eveline's home," and the preposition *in* means "at Eveline's home." The cohesive repetition of this similar deictic perspective throughout the discourse produces a sense of continuity.

Conjunctives Versus Other Semantic Elements. Halliday and Hasan's final category of semantic cohesion involves the use of conjunctives and conjunctive adverbs. For instance, consider the following two versions of a discourse:

Example 3.14a.
Pilots continually fly low-level, terrain-following missions to ranges or MOAs [military operations areas]. Because they fly at 250 knots or less, they are not required to fly on military training routes [MTRs]. The wing does not currently schedule sorties on the MTRs.

Example 3.14b.
Pilots continually fly low-level, terrain-following missions to ranges or MOAs [military operations areas]. **However,** because they fly at 250 knots or less, they

are not required to fly on military training routes [MTRs]. **Therefore**, the wing does not currently schedule sorties on the MTRs. (Dept. of the Air Force, 1991, p. 3-5; emphasis added)

Note that our perception of continuity is stronger in Example 3.14b precisely because the producer used the conjunctive elements highlighted in boldface. Note also that the cohesion principle of similarity does not explain this strengthened perception. In other words, the conjunctive adverbs do not, like other cohesive semantic elements (e.g., synonyms), produce a sense of continuity through repeated similar meanings.

Other research has noted the distinction between conjunctive elements and other semantic elements. For instance, Halliday and Hasan (1976, p. 226) argued that, unlike other semantic unifying elements, conjunctive elements are cohesive because of their specific, lexical meanings. It is important to note that some conjunctive elements do signify continuity by explicitly signaling similarity. In fact, Halliday and Hasan (1976, p. 242) designated some conjunctive elements (e.g., *likewise* and *similarly*) as additive types establishing similarity. In addition, the authors categorize conjunctive elements into four types: additives, adversatives (e.g., *however*), causals (e.g., *therefore*), and temporals (e.g., *finally*). Thus, conjunctive elements may also signify these other types of relationships between successive contributions in a discourse.

Interestingly, although conjunctive elements produce an explicit sense of discourse continuity, the other semantic elements produce that continuity implicitly. For instance, consider the following two versions of an earlier discourse example:

Example 3.15a.
For five of the items, sentences exhibiting the greatest durational distinction between morphemic and non-morphemic /s/ *were chosen.* **For the remaining three items,** two sentences with no distinction, and one with reverse distinction . . . *were selected.*

Example 3.15b.
We chose sentences exhibiting the greatest durational distinction between morphemic and non-morphemic /s/ for five of the items. **In contrast**, two sentences with no distinction, and one with reverse distinction were selected for the remaining three items.

I noted earlier that continuity is produced in Example 3.15a because of the repetition of similar syntactic structure. This background of similarity foregrounds the implicit (i.e., unstated) semantic distinction between the propositions expressed in the two sentences of the discourse. On the other hand, note that, in order to maintain the original meaning of the discourse, I had to substitute the conjunctive element, *in contrast*, rather than *similarly* in Example 3.15b.

Therefore, the conjunctive element establishes the semantic distinction between the propositions expressed in the two sentences explicitly (i.e., by stating it)—unlike syntactic parallelism, which establishes such continuity implicitly through foregrounding. Thus, within the theory of coherence and cohesion established in this book, conjunctive elements must be seen as explicit markers of continuity rather than as cohesive elements that imply continuity through their foregrounding function.

In sum, semantic elements that producers manipulate in order to produce a sense of continuity include: referents, substitutes/ellipses, lexical items, and conjunctives. However, unlike the other three categories, the unifying effect of conjunctives is not explained by the cohesion principle of similarity. Instead, conjunctive elements establish continuity explicitly by virtue of their lexical meaning.

At this point, I have covered the complete range of discourse elements that producers use as cues in order to create a sense of continuity based on similarity: visual, auditory/phonological, syntactic, and semantic. In the following section, I turn to a discussion of another sub-principle of continuity—proximity.

The Cohesion Principle of Proximity

Gestalt theory acknowledges that similarity alone is not enough to predict perceptions of continuity in visual and auditory phenomena. Lerdahl and Jackendoff (1983) used musical examples like the ones in Fig. 3.5 to illustrate the principle of proximity. Note that our most natural perception of the sounds represented in Fig. 3.5a is as one continuous whole. In contrast, despite the fact that all five notes in Fig. 3.5b display similar pitch and loudness, our most natural perception is of two wholes: a group of two notes, followed by a group of three. The principle of proximity explains these perceptions by acknowledging the impact of temporal proximity on continuity. Specifically, Fig. 3.5b is perceived as two units because the temporal space between the second and third notes is greater than that between the other notes. In fact, a visual musical symbol is inserted between the second and third notes to represent that greater temporal space. In addition, Fig. 3.5c demonstrates the gradable nature of perceptions of continuity based on proximity. Note that our perception of continuity among the three notes is stronger in Fig. 3.5c than in 3.5b precisely because of the greater temporal distance between the second and third notes. (Again, visual musical symbols represent that greater temporal distance.)

The principle of proximity may also involve spatial proximity in visual phenomena. Consider the examples below.

Example 3.16a.
■ ■ ■ ■ ■

Example 3.16b.
■ ■ ■ ■ ■

Example 3.16c.
■ ■   ■ ■ ■
(adapted from Lerdahl & Jackendoff, 1983, p. 128)

Note that our perceptions of continuity in these examples are exactly parallel to those in Fig. 3.5; one continuous whole in Example 3.16a; a group of two followed by a group of three in Example 3.16b; and an even stronger perception of two groups in Example 3.16c. In this case, however, our perceptions are based on the spatial distance between visual elements rather than the temporal distance between auditory ones.

To illustrate how the principle of proximity can explain a recipient's sense of continuity in discourse, consider the following sample discourse:

Example 3.17.
Unfortunately for Gollum, Bilbo had heard that sort of thing before; and the answer was all round **him** anyway. "Dark!" **he** said without even scratching **his** head or putting on **his** thinking cap.

> *A box without hinges, key, or lid,*
> *Yet golden treasure inside is hid,*

he asked to gain time, until **he** could think of a really hard one ... [about half a page of text occurs between these excerpts]
   But suddenly Gollum remembered thieving from nests long ago, and sitting under the river bank teaching **his** grandmother, teaching **his** grandmother to suck—"Eggses!" **he** hissed. "Eggses it is!" Then **he** asked:

> *Alive without breath,*
> *As cold as death;*
> *Never thirsty, ever drinking,*
> *All in mail never clinking* ... (Tolkien, 1966, pp. 75, 76; boldface emphasis added)

Note that all of the boldfaced, third person, masculine pronouns in the first portion of this excerpt refer to Bilbo, whereas those in the second portion refer to Gollum. Note also that few of these pronouns appear in the same clause as their referent (which would provide a syntactic explanation for our choice of referents). For those that do not, the principle of spatial proximity helps to explain our sense of the continuity of reference within each portion of the discourse.[3] In other words, the producer of this discourse used the proximity of pronouns to influence the recipients' sense of continuity within each portion of the discourse.

It is important to recognize that the principle of proximity, like that of similarity, serves a foregrounding function. If we examine Example 3.16 again,

---

[3]In fact, because recipients experience even written discourse in real-time, we might claim that both temporal and spatial proximity are involved in Example 3.17.

(a)

(b)

(c)

FIG. 3.5. A musical example demonstrating that the principle of proximity is also a foundation for the principle of continuity.

we see that it is precisely because of the equal spatial proximity that we perceive no grouping distinction in Example 3.16a. In contrast, it is the equal proximity between the last three squares in Examples 3.16b and 3.16c that foregrounds the unequal proximity between them and the first two. Turning back to Example 3.18, note that it is the proximity among the pronouns in the first portion of the excerpt and also among those in the second portion that foregrounds the greater proximity and thus the semantic distinction (i.e., the difference in referents) between the two portions of the discourse.

The Range of Cohesive Discourse Elements

The cohesive property of proximity, like that of similarity, can involve the complete range of discourse elements. In order to establish that range, the following discussion provides a few examples.

Visual Elements

For instance, consider Example 3.18 and its accompanying diagram in Fig. 3.6.

Example 3.18.
Wave height is related to wave energy and an analysis of wave height data from the hindcasting procedure indicates a higher energy regime during the fall/winter period. This conclusion is based on a plot (Fig. 3.6) of the running mean of the four highest significant wave heights for each month taken from Table [x]. This pattern of seasonal wave energy is typical of the northeastern Atlantic coast. (U.S. Army Corps of Engineers, 1990, p. 4-7)

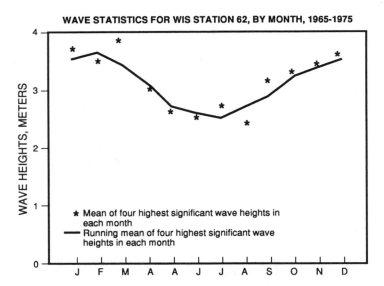

FIG. 3.6. A plot, along with related discourse (Example 3.18), illustrating the role of spatial proximity in the creation of discourse continuity.

Note that, as I'm writing this chapter, I do not know exactly where Fig. 3.6 will appear in relation to the discourse in Example 3.18. The publisher, as one of the producers of this book, will influence my readers' sense of the continuity of these discourse segments depending on how near each other they are placed. In particular, that sense of continuity will be strongest if the two elements occur next to each other, somewhat weaker if the figure appears somewhere else on the same page as the prose, and even weaker if the figure appears on a different page than the prose. Thus, the cohesion principle of proximity is gradable and can involve visual as well as semantic elements.

Auditory/Phonological Elements

Obviously such devices as rhyme and alliteration require the proximity of phonological elements. For instance, consider the following two versions of a discourse:

Example 3.19a.
Not in a box.
Not with a fox.
Not in a house.
Not with a mouse.
I would not eat them here or there.
I would not eat them anywhere.
I would not eat green eggs and ham.
I do not like them. Sam-I-am.

Example 3.19b.
Not in a box.
Not in a house.
I would not eat them here or there.
I would not eat green eggs and ham.
Not with a fox.
Not with a mouse.
I would not eat them anywhere.
I do not like them Sam-I-am. (adapted from Seuss, 1960, p. 24)

Note that the producer of Example 3.19a has established a strong sense of continuity, partly through the use of rhyme. In contrast, although the same sentences are used in Example 3.19b the sense of continuity is significantly weaker. In fact, I am not sure I could even claim that Example 3.19b does rhyme precisely because of the lack of sufficient proximity between the relevant words. Thus, similarity alone is not enough to produce a strong sense of continuity in Example 3.19b. I explore this relationship between similarity and proximity further in chapter 4. But, first, let's consider how the cohesion principle of proximity applies to other discourse elements.

Semantic Elements

I noted the cohesive effect of proximity involving reference in Example 3.17. Other types of semantic elements may also produce a sense of continuity based on proximity. For instance, conjunctive elements can explicitly signify the proximity as well as the similarity or dissimilarity of propositions expressed in a discourse. As one example, consider the following discourse excerpt attributed to Lewis Carroll:

Example 3.20.
"There's no sort of use in knocking," said the Footman, "and that for two reasons. **First**, because I'm on the same side of the door as you are; **secondly**, because they're making such a noise inside, no one could possible hear you." (Halliday and Hasan, 1976, p. 264; emphasis added)

Note that the boldfaced conjunctive elements explicitly state the temporal relationship of the propositions expressed by the two clauses in the second sentence to the proposition expressed in the first. As I noted earlier, one of Halliday and Hasan's four categories of conjunctive elements is labeled "temporal," and they list a host of examples (e.g., *previously* and *meanwhile*). In all cases, however, I have chosen not to label conjunctive elements as **cohesive** because they do not produce continuity implicitly through a foregrounding function. Instead, they function as an explicit statement of the continuity of the discourse.

In order to further emphasize the distinctiveness of conjunctive elements, consider the following discourse:

Example 3.21.
First, consider some questions concerning X-bar theory. . . . Consider **next** the
theory of movement. . . . **To review**, this discussion has touched on issues related
to various components of UG, particularly the theories of government and bounding.
(adapted from Chomsky, 1986, pp. 2, 4, 87; emphasis added)

Note that, although all three boldfaced conjunctive elements affect perceptions
of coherence, only the first two (i.e., *first* and *next*) establish continuity. In
contrast, the third conjunctive element (i.e., *to review*) establishes coherence
based on quantity of information or completeness.

At this point, we have seen that proximity is another cohesive principle that
allows producers to establish continuity in discourse. In addition, I have noted
that, like the cohesion principle of similarity, the cohesion principle of proximity
is gradable and may involve a range of discourse elements. Moreover, I have
shown that some conjunctive elements explicitly signify a proximity relation
between discourse contributions.

SUMMARY

In this chapter, I discussed two cohesion principles that establish discourse
continuity. In brief, I argued that discourse producers influence recipients' sense
of discourse continuity by manipulating the similarity and proximity of the full
range of discourse elements. The cohesion principle of similarity acknowledges
the cohesive effect of similar discourse elements, whereas the cohesion principle
of proximity acknowledges the effect of the spatial and temporal proximity of
discourse elements. In addition, I noted that all cohesion serves to highlight
semantic distinctions in discourse by providing a homogenous background against
which those distinctions are foregrounded. In contrast, I noted two qualities of
conjunctive elements that caused us to label them as markers of coherence rather
than as markers of cohesion: (a) they express coherence explicitly by virtue of
their lexical meanings rather than implicitly through a foregrounding function
like the cohesion principles, and (b) they express continuity as well as other
types of discourse coherence (e.g., related to the coherence principle of quantity).

Furthermore, this proposed theory of cohesion accounts for the impact of the
full range of discourse elements: visual, auditory/phonological, syntactic, and
semantic. Moreover, I showed that this theory might provide an explanation of
the relationship between cohesion and coherence: cohesion influences one type
of coherence—continuity. I explore this relationship further in chapter 5.

Global Cohesion and Discourse Organization: The Interaction of Cohesive Elements

In chapter 2, I established a theory of coherence comprised of four principles that describe the qualities that recipients attribute to coherent discourse: continuity (relation), clarity (manner), adequacy (quantity), and accuracy (quality). In chapter 3, I established a theory of cohesion comprised of two principles that describe how producers establish one aspect of discourse coherence—continuity; those principles acknowledge the impact of the similarity and proximity of discourse elements. However, I have considered how continuity is established only at the most local or lowest level of discourse—what I have called discourse elements (e.g., typographies, phonemes, morphemes, clauses, words, etc.).

In this chapter, I explore the question of how cohesion principles interact with each other to create organization in discourse. First, I introduce two meta-principles based on Gestalt theory that describe how individual cohesive elements interact; these include the principles of reinforcement and conflict. Second, I discuss the role of local cohesive elements in producing higher level organization or global cohesion within a discourse, and expand the theory to include two more principles based on Gestalt theory; these include the principles of intensity, and of size and symmetry.

THE METAPRINCIPLES OF REINFORCEMENT AND CONFLICT

Gestalt theory recognizes the combined impact of principles such as similarity and proximity on our sense of unity by proposing two further principles: reinforcement and conflict. To begin, consider the following visual examples:

Example 4.1a.
■ ■ ■ ■ ■

Example 4.1b.
■ ■ ■ ■ ■

Example 4.1c.
■ ■ ■ ■ ■

Example 4.1d.
■ ■ ■ ■ ■

Example 4.1e.
■ ■ ■ ■ ■

Note that we have seen examples similar to 4.1a, 4.1b, and 4.1c in chapter 3. I noted then that Example 4.1a is most naturally perceived as one continuous whole; 4.1b as one group of two followed by one group of three based on the principle of similarity; and 4.1c with the same arrangement of 4.1b based on the principle of proximity.

In addition, I can now note that Example 4.1d is a kind of combination of Examples 4.1b and 4.1c in which we perceive the same arrangement of a group of two followed by a group of three even more strongly. The principle of reinforcement explains the strength of this perception by acknowledging the impact of the interaction of similarity and proximity when they promote the same perception. In contrast, note that Example 4.1e is perceived as ambiguous. We see two equally natural units here: (a) one group of two followed by a group of three (based on similarity), or (b) one group of three followed by a group of two (based on proximity). The principle of conflict explains the ambiguity of this perception by acknowledging the impact of the interaction of similarity and proximity when they promote different perceptions.

Again, as the following musical examples show in Fig. 4.1, we perceive auditory phenomena according to the same principles. Note that our perceptions of the sounds represented in Fig. 4.1 are parallel to those of the visual examples (4.1): Fig. 4.1a as one continuous whole; Fig. 4.1b as two groups of notes based on similarity in pitch; Fig. 4.2c as two groups of notes based on proximity or rhythm; Fig. 4.1d as two groups of notes more strongly perceived based on reinforcement of similarity and proximity; and, finally, Fig. 4.1e as ambiguous based on the conflict between similarity and proximity.

As an introductory example of how the principles of reinforcement and conflict explain recipients' perceptions of discourse continuity, consider our previous example from *The Hobbit*, which is repeated here for convenience:

Example 4.2.
Unfortunately for Gollum, Bilbo had heard that sort of thing before; and the answer was all round **him** anyway. "Dark!" **he** said without even scratching **his** head or putting on **his** thinking cap.

A box without hinges, key, or lid,
Yet golden treasure inside is hid,

he asked to gain time, until **he** could think of a really hard one ... [about half a page of text occurs between these excerpts]

But suddenly Gollum remembered thieving from nests long ago, and sitting under the river bank teaching **his** grandmother, teaching **his** grandmother to suck—"Eggses!" **he** hissed. "Eggses it is!" Then **he** asked:

Alive without breath,
As cold as death;
Never thirsty, ever drinking,
All in mail never clinking.... (Tolkien, 1966, pp. 75, 76; boldface emphasis added)

I noted in chapter 3 that cohesive visual elements (e.g. italic typography) produce a sense of continuity (through the principle of similarity) among the last four lines. In addition, note that the same visual elements occur in two earlier lines in this excerpt, thus creating a sense of continuity among all six italicized lines (i.e., they comprise riddles in the discourse). Furthermore, I noted in chapter 2

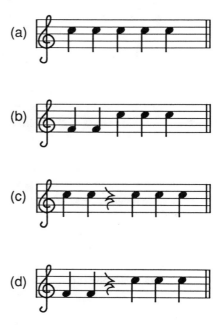

FIG. 4.1. Musical examples illustrating the interaction of cohesion principles as they apply in auditory phenomena.

that visual elements can produce a sense of continuity through the principle of proximity. Thus, we saw that the proximity of the boldfaced pronouns to different referents created a sense of continuity among the first portion of the discourse and among the second portion, but a sense of discontinuity between the portions (i.e., Bilbo's riddle vs. Gollum's riddle).

Therefore, this discourse excerpt demonstrates each of the principles noted in the visual and auditory examples introduced earlier. First, the cohesion principle of similarity explains how the repetition of similar typography prompts a recipient's perception of continuity between the excerpts. Second, the principle of reinforcement explains how this repetition of typography in combination with the repetition of a similar text type or genre (i.e., the riddle) prompts a recipient's even stronger perception of continuity between the excerpts. Third, the principle of proximity explains how the proximity of pronoun referents prompts a recipients' perception of the lack of continuity between the excerpts (i.e., Bilbo vs. Gollum). And, finally, the principle of conflict explains how both the similarity (i.e., typography and text type) and the proximity (i.e., referents) of discourse elements produce a recipient's simultaneous perception of both the continuity and the lack of continuity in this discourse example. In other words, these principles explain how recipients perceive two, distinct discourse segments that are nevertheless related.

THE INTERACTION OF LOCAL COHESIVE ELEMENTS

As an example of how the local principles of similarity and proximity work to reinforce or ambiguate our perceptions of discourse unity, consider the following discourse sample made up of sentences from one section in a medical case report:

Example 4.3a.

#1: On **physical examination**, the foal was thin, depressed, and febrile (39.4 C), with an elevated heart rate (120 beats/min) and respiration rate (48 breaths/min)....

#2: The **hemogram** revealed hypoproteinemia (4.8 g/dl), a marked leukocytosis (94,200/ul), and mature neutrophilia (91,400/ul)....

#3: **Thoracic radiographs** revealed widespread pulmonary abscessation....

#4: Another **biochemistry profile** showed that ALP (1180 U/L) had continued to rise. . . .

#5: **Ultrasonographic imaging** of the liver was normal.... (Freestone, Shoemaker, & McClure, 1990)

Note that each of these discourse excerpts in Example 4.3 appears in paragraph-initial position (i.e., the same discourse position). Note also that the boldfaced terms are all semantically similar. Specifically, they all function in the same semantic case role: as **instruments**. In other words, *physical examination, hemogram, thoracic radiographs, biochemistry profile*, and *ultrasonographic* *imaging* are noun phrases expressing the instruments by which the examiner noted the horse's condition. In addition, they are all hyponyms (i.e., types) of "medical, diagnostic aids." Thus, the sense of continuity throughout the discourse section containing the paragraphs in which these sentences occur is explained by the principle of reinforcement, which acknowledges the impact of these two kinds of similarities (in placement or arrangement and in semantic relation), both of which prompt the same perception of continuity among these excerpts. In addition, note that these sentences occur in consecutive paragraphs. Thus, the principle of proximity reinforces the same sense of continuity established by the principle of similarity.

Now consider the difference in a recipient's perception of the continuity of this section of the case report in the following altered version of Example 4.3a:

Example 4.3b.

#1: On **physical examination**, the foal was thin, depressed, and febrile (39.4 C), with an elevated heart rate (120 beats/min) and respiration rate (48 breaths/min)....

#2: The **hemogram** revealed hypoproteinemia (4.8 g/dl), a marked leukocytosis (94,200/ul), and mature neutrophilia (91,400/ul)....

#3: **Thoracic radiographs** revealed widespread pulmonary abscessation....

#4: Differentials for icterus in the horse include anorexia, hemolytic anemia, liver disease and cholestasis....

#5: Chenodeoxycholic acid (CDCA) is a bile acid used in human patients for the dissolution of bile stones....

#6: Another **biochemistry profile** showed that ALP (1180 U/L) had continued to rise. . . .

#7: **Ultrasonographic imaging** of the liver was normal....

In this version, the producer has placed two intervening paragraphs without similar case roles/hyponyms in their paragraph-initial sentences (i.e., #4, #5) between those paragraphs with similar hyponyms in their initial sentences (i.e., #1, #2, #3, #6, #7). In this case, the principle of proximity would not produce the same sense of continuity in the discourse section as the principle of similarity, and, therefore, the principle of conflict would explain the recipient's ambiguous sense of continuity in this discourse section. For example, proximity suggests continuity between paragraphs beginning with #3 and #4, whereas similarity suggests continuity between paragraphs beginning with #3 and #6. Eventually, if we increased the distance between the paragraphs beginning with #3 and #6, the perception based on proximity would win out over the perception based on similarity. Thus, the ambiguous sense of both continuity and lack of continuity in this entire discourse section would be replaced by an unambiguous sense of lack of continuity.

Interestingly, the cohesion principles of similarity and proximity appear to be the foundation for the widely documented cohesive/coherence theory, functional sentence perspective (FSP), also known as the given-new contract (e.g., see Daneš, 1974). To illustrate, compare the following three versions of a discourse:

Example 4.4a.
An excellent example of an <u>epic poem</u> is The Odyssey. A long narrative or story is usually included in *epic poems*. Certain conventions almost always mark this story.... [underlined emphasis added]

Example 4.4b
The Odyssey is an excellent example of an <u>epic poem</u>. *Epic poems* usually include a long narrative or story. This story is almost always marked by certain conventions. ... [underlined emphasis added]

Example 4.4c
The Odyssey is an excellent example of an epic poem. **The Odyssey** includes a long narrative or story. **Homer's work** is also marked by certain conventions.... (adapted from Vande Kopple, 1982, pp. 53–54; emphasis added)

Note that Example 4.4a lacks continuity, whereas 4.4b and 4.4c do not. More importantly, note that both Examples 4.4a and 4.4b demonstrate the repetition of lexical items (e.g., the underlined and italicized phrase)—one kind of semantic cohesion. However, note that the proximity of the repeated items is much greater in Example 4.4a than in Example 4.4b. The closer proximity of the items in Example 4.4b is what accounts for the greater continuity in that example. In terms of FSP, Example 4.4b follows the AB–BC arrangement of information in which new information (e.g., the underlined phrase in Example 4.4b) designated as "B" in a preceding sentence becomes the given information (e.g., the italicized phrase) also designated as "B" in the following sentence. Because new information typically appears near the end of the sentence and given information near the beginning, the proximity of cohesive semantic elements is the foundation of the AB–BC pattern.

In addition, note that Example 4.4c also demonstrates semantic cohesion (e.g., in the boldfaced terms) through the repetition of lexical items and the use of synonymy (i.e., *Homer's work*). However, the greater similarity in Example 4.4c is what distinguishes it from the problematic version in Example 4.4a. In terms of FSP, Example 4.4c follows the AB–AC arrangement of information in which the given information (e.g., the boldfaced phrases in Example 4.4c) designated as "A" in each sentence remains the same. Again, because given information typically appears near the beginning of the sentence, the similar syntactic placement of the similar semantic elements are the foundation of the AB–AC pattern.

These examples demonstrate that, when the similarity and proximity of discourse elements reinforce each other, they explain the greater sense of continuity in discourse. It is important to note that the principle of similarity applies to semantic elements in cases of either AB–BC or AB–AC arrangement. Obviously, this semantic cohesion is crucial in establishing continuity. However, FSP acknowledges that cohesion is enhanced when producers reinforce this semantic connection with other cohesive elements. In the case of the AB–BC pattern, semantic similarities are reinforced by their proximity. And, in the case of the AB–AC pattern, semantic similarities are reinforced by their similar sentence position.

Like FSP, the theoretical construct of discourse focus is also explained by the cohesion principles of similarity and proximity. Focus is used to describe one aspect of discourse connectedness in computational linguistics and is clearly related to some early research in FSP (see Daneš, 1974). For example, consider the following brief discourse:

Example 4.5.
Put the mud pack on your face.
After 5 minutes, pull it off. (Sidner, 1983, pp. 318)

Note that *the mud pack* is the focus or topic of the first contribution (signaled by the definite article *the*) and that it is maintained as the topic in the second contribution by the referent or anaphor *it*. Note also that Example 4.5 follows the AB–AC pattern described by FSP theory: the same given information appears in the same syntactic spot (i.e., as subject) in each contribution. Therefore, in the terms of our theory, continuity is established in this discourse by the use of a type of semantic cohesion, repetition of a semantic element through the use of a referential term, and is reinforced by the similar syntactic placement of those related semantic elements as well as their proximity to each other.

In the terms of focus theory, connectedness is established through the focus on one topic—*the mud pack*—in this discourse by the use of an anaphoric term appearing in a following, consecutive contribution. Thus, both theories provide similar accounts of the continuity in Example 4.5 in which repetition of similar elements and their proximity to each other play the crucial role. However, focus theory attends only to one sub-type of semantic cohesion—anaphora or reference. Sidner (1983) recognized that the semantic connection between the two contributions in Example 4.5 is reinforced by the "similarity of the syntactic structure of the two sentences, each containing an imperative mood main clause" (p. 318), but states that examples like 4.6, in which syntactic similarity conflicts with semantic similarity, are problematic for focus theory:

Example 4.6.
The green Whitierleaf is most commonly found near the wild rose.
The wild violet is found near it too. (Sidner, 1983, p. 318)

Note that the use of the referential *it* in the second sentence creates a sense of continuity in this short discourse. In addition, note that a number of lexical items are repeated in the two sentences and that the syntactic structure of both sentences is also quite similar.

In terms of focus theory, *Whitierleaf* is clearly the focus after the first sentence in Example 4.6. However, Sidner claimed that *it* refers back to *rose*, thus changing the focus in the second sentence. In contrast, resolving the identity of the original referent for *it* appears to be ambiguous: either *rose* or *Whitierleaf* is equally plausible. In terms of this theory, this ambiguity is explained by noting that the recipient's sense of continuity based on the similarity of syntactic structure (which suggests that *Whitierleaf* is the referent) conflicts with his or her sense of continuity based on the proximity of the two possible referents (which suggests that *rose* is the referent). Sidner (1983) argued that "between similarity of [syntactic] structure and the focus rules, similarity is preferred as a means of choosing the [original referent], so when each gives a different prediction, similarity of structure must be used" (p. 318). Despite the need to consider the interaction of local cohesive elements in order to predict a recipient's sense of continuity, Sidner noted that research has not yet provided computational means for accurately recognizing syntactic parallelism.

In this section, I have shown that, when the local cohesion principles of similarity and proximity reinforce the same sense of continuity, the recipient's sense of continuity is stronger. In contrast, I have noted that, when these local principles conflict with each other, the recipient's sense of continuity is weaker and sometimes ambiguous. In addition, I have noted that the theory of FSP or the given-new contract and the theory of focusing are founded on the meta-principle of reinforcement in which semantic cohesion is reinforced through local cohesion created by similarity and/or proximity.

The Foregrounding Function of Similarity

I noted in chapter 3 that the cohesion principles of similarity and proximity produce a sense of continuity. I have argued elsewhere that the repetition of similar structural discourse elements is important in producing a sense of continuity because the repetitions serve as a background against which dissimilar semantic elements are foregrounded (Campbell, 1991, 1992). I used Fig. 4.2 as a visual example. Try to locate the *O* in both visual arrangements. Although each *O* appears in the same row and column of each example, it is significantly easier to locate the *O* in Fig. 4.2 (1b) because the repetition of *X*s (which are in close spatial proximity) produces a uniform background against which the *O* is foregrounded. Thus, the cohesion principles of similarity and proximity necessarily describe not only continuity, but also discontinuity.

To illustrate how these cohesion principles apply in discourse, consider the following example from the *MS-DOS User's Reference* (1987) manual:

Example 4.7.
The **find** command looks for *string* in one or more files. After searching the specified files, **find** displays any lines it has found that contain the specified string. (Microsoft Corporation, p. 72)

```
(1a)  XTELPCLWA        (1b)  XXXXXXXXX
      BCZNRFILQ              XXXXXXXXX
      KPWSATCVI             XXXXXXXXX
      MCHLPWODU             XXXXXXOXX
      ISCQPYDZL             XXXXXXXXX
```

FIG. 4.2.   A visual example illustrating the foregrounding function of proximate repetitions of similar visual features.

Note that the word *string* appears in both sentences in Example 4.7. The principle of similarity explains the sense of continuity between the two sentences by acknowledging the impact of this repetition. However, in the first sentence the word occurs in italics, whereas in the second it occurs in roman typography. The similarity of the two words foregrounds their dissimilar typography, and thus suggests a distinction in their meaning.

As another example of the foregrounding function of proximate similarities in discourse elements, consider the bulleted portion of the discourse example from chapter 3, which is repeated here:

Example 4.8.
Mr. Krishan Saigal, P.E., will serve as Lead Engineer. Mr. Saigal's primary tasks will include:

• Plan and provide direction for technical work elements.

• Coordinate technical direction of subcontractors.

• Assist in coordinating and disseminating project-related information to the Project Team. . . .

Mr. Saigal will also serve as Construction Manager for the Project Team, with the following primary responsibilities. . . . (SCS Engineers, 1991, p. 2-2)

I noted earlier that the cohesion principle of similarity accounts for the effect of the repetition of similar propositional content, syntactic structure, and visual cues in establishing a sense of continuity. Note also that these similar discourse elements are placed in close spatial proximity. Thus the repetition of similar, proximate discourse elements produces a background against which the semantic distinctions within this bracketed segment are foregrounded and, therefore, are more likely to be noticed by the recipient (e.g., one foregrounded distinction involves the types of activities the engineer will carry out—*planning, coordinating,* and *assisting*). Moreover, the similar, proximate discourse elements create a sense of continuity among the three items, which foregrounds the discontinuity before and after the bracketed portion of the discourse in Example 4.8.

Interestingly, discontinuity appears to be the crucial factor in creating global organization within a discourse. The general, cognitive principles that direct such organization are the subject of some previous research in music theory. Lerdahl and Jackendoff (1983) provided an insightful study of tonal music in which they propose that Gestalt theory describes the cognitive principles that direct how listeners cognitively organize or structure tonal music into segments. As the authors noted:

> [I]t is evident that a listener perceives music as more than a mere sequence of notes with different pitches and durations: one hears music in organized patterns. Each rule of musical grammar is intended to express a generalization about the organization that the listener attributes to the music he hears. (p. xii)

Clearly, recipients also perceive discourse as "more than a mere sequence" or as more than one continuous stream of sounds or graphemes or even sentences. In fact, Halliday and Hasan (1976) noted the importance of lack of cohesion or discontinuity in signaling discourse organization:

> [I]f a sentence shows no cohesion with what has gone before, this does indicate a transition of some kind; for example, a transition between different stages in a complex transaction, or between narration and description in a passage of prose fiction. We might choose to regard such instances as discontinuities, signaling the beginning of a new text. (p. 295)

Thus, continuity and discontinuity appear to be relevant at multiple levels in discourse. At the most global level (i.e., the level of the whole discourse), I have used principles of coherence to explain a recipient's sense of continuity. At the most local level (i.e., the level of discourse elements: e.g., morphemes, sentences, typography, etc.), I have used principles of cohesion to explain how a producer influences a recipient's sense of continuity. And, finally, at the level in between (i.e., the level of sections, paragraphs, etc.), I use global cohesion principles to explain how the local level cohesive elements influence a recipient's sense of discontinuity in discourse.

In the next section of chapter 4, I introduce two Gestalt principles involved in establishing discontinuity: intensity, and size and symmetry. In addition, I argue that, like the principles of similarity and proximity, these organizing principles describe a type of cohesion. However, unlike the other cohesion principles, these organizing principles describe cohesion that occurs at a higher or more global level of discourse.

The Global Cohesion Principle of Intensity

The Gestalt principle of intensity acknowledges the impact of the intensity of the similarity or proximity of visual or auditory phenomena on perceptions of unity. Lerdahl and Jackendoff (1983) used a musical example similar to the one in Fig. 4.3 in order to illustrate. Note that listeners perceive a number of breaks or discontinuities in this musical example that divide the example into segments as indicated with brackets. The perception of five individual segments at the lowest level (a) within this example is due to the temporal proximity of the notes as signaled in the musical score by the symbol ⌣. However, the perception of two segments at the higher level (b) within this example is due not just to temporal proximity, but to the degree or intensity of that proximity between notes 9 and 10 as signaled in the score by the symbol ⋛. Thus, the principle of intensity explains the organization of a musical piece into higher level segments based on the relative degree of the proximity or similarity of the individual auditory elements (i.e., notes, in this case). This example clearly demonstrates how the principle of intensity creates discontinuity that is, in turn, the foundation for setting boundaries that subdivide the stream of auditory elements into more global segments.

To illustrate how the principle of intensity explains some of a recipient's sense of discontinuity and organization in discourse, consider the memo in Fig. 4.4. Note that recipients of this discourse are likely to sense a break or discontinuity between lines 8 and 9. In terms of the theory established so far, the discontinuity between lines 8 and 9 is the result of the relative lack of spatial proximity or of cohesion established through the principle of proximity between these lines (e.g., compared to lines 7 and 8 or lines 9 and 10). In simpler terms, lines 1 through 8 comprise one continuous segment of the discourse for which we even have a conventional label (i.e., *heading*), whereas lines 9–30 comprise another continuous segment of the discourse for which we also have a conventional label (i.e., *body*).

Note that recipients also sense discontinuity between the lines numbered 13 and 14 because of the relative lack of spatial proximity (e.g., compared to lines 14 and 15). But, more importantly to my purpose here, note that recipients sense more discontinuity in the case of 8/9 than they do in the case of 13/14. The principle of intensity is one explanation for this observation. In particular, because the principle of proximity applies less intensely to lines 8/9 than it does to 13/14 (i.e., 8/9 are less proximate than 13/14), recipients sense less continuity/more discontinuity between lines 8 and 9.

I have noted repeatedly that the cohesion and coherence principles are gradable. Fig. 4.4 makes it clear that the principle of intensity acknowledges the impact of

FIG. 4.3. A musical example demonstrating the influence of intensity on perceptions of continuity in auditory phenomena.

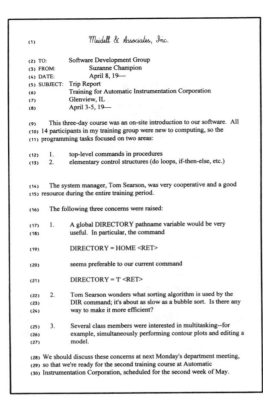

(1)   *Meidell & Associates, Inc.*

(2) TO: Software Development Group
(3) FROM: Suzanne Champion
(4) DATE: April 8, 19—
(5) SUBJECT: Trip Report
(6)   Training for Automatic Instrumentation Corporation
(7)   Glenview, IL
(8)   April 3-5, 19—

(9)   This three-day course was an on-site introduction to our software. All
(10) 14 participants in my training group were new to computing, so the
(11) programming tasks focused on two areas:

(12)   1.   top-level commands in procedures
(13)   2.   elementary control structures (do loops, if-then-else, etc.)

(14)   The system manager, Tom Searson, was very cooperative and a good
(15) resource during the entire training period.

(16)   The following three concerns were raised:

(17)   1.   A global DIRECTORY pathname variable would be very
(18)   useful. In particular, the command

(19)   DIRECTORY = HOME <RET>

(20)   seems preferable to our current command

(21)   DIRECTORY = 'T' <RET>

(22)   2.   Tom Searson wonders what sorting algorithm is used by the
(23)   DIR command; it's about as slow as a bubble sort. Is there any
(24)   way to make it more efficient?

(25)   3.   Several class members were interested in multitasking--for
(26)   example, simultaneously performing contour plots and editing a
(27)   model.

(28) We should discuss these concerns at next Monday's department meeting,
(29) so that we're ready for the second training course at Automatic
(30) Instrumentation Corporation, scheduled for the second week of May.

FIG. 4.4. A complete discourse example used to illustrate the results of the interaction of individual, local cohesive elements on perceptions of discourse organization. From *Technical Communication, Second Edition* (p. 452, Example 15.3) by Rebecca Burnett, 1990, Belmont, CA: Wadsworth Publishing Co. Copyright © 1990 by Wadsworth Publishing Co. Reprinted by permission.

the relative degree of cohesion on a recipient's sense of discontinuity. In this case, the more white space between the lines, the greater a recipient's sense of discontinuity would be. Once again, we must recognize the foregrounding function of cohesion. In Fig. 4.4, we can note that it is precisely because of the nearer spatial proximity between lines 13 and 14 that the greater spatial proximity between lines 8 and 9 is foregrounded.

Let's consider another example in which the principle of intensity applies to the similarity rather than proximity of discourse elements. For instance, note that a recipient senses greater discontinuity between the phrase expressed in line 13 and the sentence expressed in lines 14–15 (i.e., one of the boundaries of the first paragraph) than he or she does between the sentence expressed in lines 25–27 and the sentence expressed in lines 28–30 (i.e., one of the boundaries of the third enumerated item within the third paragraph). However, note that the intensity of spatial proximity cannot explain this observation because the proximity of lines 13 and 14 is equivalent to that of lines 27 and 28.

Instead, it appears that the degree of similarity between lines 13 and 14–15 is less than between lines 25–27 and 29–30. In both cases, the producer has established some semantic cohesion through the principle of similarity. On the one hand, note that lines 13 and 14–15 use semantically related terms (e.g., *control structures* and *system managers* are collocationally related in that they both deal with the vocabulary of computers). On the other hand, note that lines 25–27 and 28–30 use more closely related terms (e.g., all of lines 25–27 is used as the (partial) referent for *these concerns* in line 28). Thus, the principle of intensity explains why the lesser degree of semantic similarity or cohesion between lines 13 and 14–15 compared to lines 25–27 and 28–30 results in a recipient's sense of greater discontinuity in 13/14–15.

In the case of intensity involving both similarity and proximity, a recipient's sense of greater discontinuity at a point in the discourse signals the boundary of a higher level or more global discourse segment. For instance, consider again the sense of discontinuity based on the intensity of the proximity of lines 8/9 compared to 13/14. In particular, the greater sense of discontinuity between lines 8 and 9 signals the boundary of a higher level discourse segment: in this case between the heading and the body (i.e., the two highest level segments of this discourse). In contrast, the lesser sense of discontinuity between lines 13 and 14 signals a boundary between lower level discourse segments: in this case between the first and second segments (i.e., two lower level segments or paragraphs) within the body of this discourse.

Similarly, consider again the sense of discontinuity based on the intensity of similarities of sentences in lines 13/14–15 compared to those in lines 25–27/28–30. On the one hand, the lesser sense of discontinuity between lines 25–27 and 28–30 signals a boundary in a lower level discourse segment: in this case between the three enumerated segments and the remainder of the third segment of the body in which all of the enumerated segments occur. On the other hand, I just noted that the discontinuity between lines 13 and 14–15 signals a boundary between the first and second segments of the body. But compared to lines 25–27/28–30, the sense of discontinuity in 13/14–15 is greater, thus signaling a boundary in a higher level discourse segment (i.e., between segments or paragraphs within the body vs. between two segments of a segment or paragraph within the body). The main point here is that the principle of intensity describes how discontinuities affect the segmenting or organization of a discourse.

To sum up, I have shown that the principle of intensity describes the impact of the relative degree of similarity and proximity of discourse elements on a recipient's sense of discontinuity within a discourse. Specifically, I have noted that the less intense the similarity or the less near the proximity of individual discourse elements, the greater the recipient's sense of discontinuity. Thus, the producer influences a recipient's sense of discourse organization by establishing discontinuity in cohesive elements. That discontinuity, in turn, signals the boundaries of segments within a discourse, with higher level discourse segments defined by the greater degrees of discontinuity and vice versa.

The Global Cohesion Principle of Size and Symmetry

If we consider the memo in Fig. 4.4 again, we see that there are a number of possible ways in which discontinuities might establish boundaries between the highest level segments of the body of this discourse (i.e., lines 9–30). Gestalt theory predicts that all of my readers sensed some particular organization when they first read the memo. However, the theory also predicts that not all of my readers sensed exactly the same organization. I think the most likely possibilities include: (a) three segments comprised of lines 9–13, lines 14–15, and lines 16–30; (b) two segments comprised of lines 9–15 and lines 16–30; or (c) two segments comprised of lines 9–13 and lines 14–30.

Unfortunately, the global cohesion principle of intensity does not provide a way to describe why a recipient chooses one over the other of these multiple possible organizations. First, the discontinuities predicted by the intensity of the proximity are equivalent in all cases: there is the same amount of white space between lines 13 and 14 as between lines 15 and 16. Second, the discontinuities predicted by the intensity of the similarities appear equivalent to me in all cases (i.e., the segment comprised of lines 14–15 seems equally disjoint from the preceding and the following discourse segments). Therefore, we have no explanation for why my readers and I have an opinion about the way in which the body of this memo is segmented or organized.

The solution in this case lies in a combination of Gestalt principles used by Lerdahl and Jackendoff (1983) to explain how music listeners decide between possible alternative structures or organizations in a musical piece. The authors used an example like the one in Fig. 4.5 in order to demonstrate. If listeners perceive segments when they hear the 12 notes represented in Fig. 4.5, the organization of that segmenting might include two possibilities: (a) two segments comprised of notes 1–3 and notes 4–12; or (b) two segments comprised of notes 1–6 and notes 7–12. Like the possible organizations of the discourse example noted earlier, the fact that a listener naturally chooses one of these possible organizations of a musical example cannot be explained by reference to the principle of intensity because all discontinuities that might signal segment boundaries reflect the same degree of proximity and similarity in their individual elements. In other words, both organizations reflect the same degree of temporal proximity (i.e., timing or rhythm) and the same degree of similarity (i.e., pitch, stress, etc.).

Lerdahl and Jackendoff (1983) claimed that the most likely perception of the organization of Fig. 4.5 is that represented by (b): two segments comprised of notes 1–6 and notes 7–12. The authors ruled out the other possible organization (b) based on two, related principles: size and symmetry. Specifically, where the

FIG. 4.5. A musical example demonstrating the role of the principles of size and of symmetry on perceptions of organization among equally plausible alternatives.

principle of intensity does not determine one possible organization, listeners prefer to subdivide musical pieces into equal or symmetrical segments. In addition, listeners prefer not to create very small segments. Thus, the principle of size and symmetry explains why listeners choose to organize the musical example in Fig. 4.5 into the two segments of equal, symmetrical size (arrangement b) rather than into two segments of unequal, asymmetrical size (arrangement a).

As an introductory example of the way in which the principle of size and symmetry can explain how global cohesion is achieved, consider the three possible organizations posited for the body of the memo in Fig. 4.4 again. First, lines 14–15 can stand alone as one of three segments in the body of the memo; in fact, the five-space indention is one page-arrangement cue that lines 14–15 comprise one of three segments or paragraphs. Thus, the producer of this memo has encouraged perception of this arrangement. However, this arrangement ignores the principle of size, which acknowledges a recipient's preference against very small discourse segments. In particular, this segment would be comprised of one, short sentence, but would function as a relatively high-level segment in the discourse (i.e., one of three segments within the body). As a recipient, I do not find this subdivision completely satisfying despite the cues (i.e., conventionally dictated five-space indention for paragraphs) placed in the discourse by the producer.

Next, consider that lines 14–15 can be combined with the preceding lines (9–13) or the following lines (16–30) to form one of two segments in the body of the memo. Both of these arrangements satisfy the principle of size because lines 14–15 function as part of a larger segment in both cases. However, the latter arrangement ignores the principle of symmetry. In particular, this segmenting of the body of the memo results in an unbalanced or unequal division of the body because lines 14–30 comprise a much larger segment than their "sister" segment (lines 9–13). In contrast, the former arrangement in which lines 9–15 function as the first segment of the body satisfies both the principle of size and of symmetry. Specifically, this arrangement of the memo results in two segments, neither of which is extremely small, and which are more equal or symmetrical in size than the previous arrangement. I cannot judge for my readers, but this latter arrangement is the one

I see as most intuitively satisfying. Thus, the principles of size and symmetry explain my sense of the organization or segmenting of the discourse in Fig. 4.4.

As another example, consider the following conversation:

Example 4.9.

CHAD	I go out **a lot.**
DEBORAH	I go out and eat.
PETER	You go out?
	The trouble with ME is
	if I don't prepare
	and eat well,
	I eat **a LOT** . . .
	Because **it's not** satisfying.
	And so if I'm just eating like cheese and crackers,
	I'll just STUFF myself on cheese and crackers . . .
DEBORAH	Oh yeah?
PETER	I've noticed that, yeah.
DEBORAH	Hmmm . . .
	Well then it works,
	then it's a good idea.
PETER	It's a good idea in terms of eating,
	it's not a good idea in terms of time. (Tannen, 1989, p. 75)

Tannen noted that the boldfaced terms are repetitive. According to the theory established here, these similarities establish local cohesion within this discourse excerpt. Interestingly, the principle of size and symmetry may explain why Tannen chose to present this particular excerpt of the conversation she analyzed. More specifically, it seems likely that the two sets of repetitive sections neatly defined by their symmetrical shape and size (i.e., each segment begins and ends with a repeated phrase) and their "not too small" size influenced Tannen's choice about where to segment the whole conversation with which she was working. The most important point here is that, among other things, the principle of size and symmetry predicts that if the two repetitions of *it's not* were found in two consecutive lines, recipients would be less likely to see them as the boundaries of an equivalent segment to the first, which is defined by the two repetitions (i.e., *I x a lot*).

SUMMARY

In this chapter, I have shown that the global cohesion principles of intensity and of size and symmetry help to explain how a recipient's sense of discourse organization is established. The principle of intensity acknowledges the impact of the relative degree of similarity and proximity of discourse elements on a recipient's sense of discontinuity. I explained that the less intensely these local cohesion principles apply, the more discontinuity is created, which, in turn, causes recipients to organize discourse into continuous segments. Because discontinuities usually suggest multiple possibilities for segmenting discourse, the principle of size and symmetry provides a way of predicting which of these possible organizations for segmenting discourse is most likely to be preferred by recipients. I noted that recipients avoid creating very small discourse segments and that they prefer to subdivide a discourse into symmetrical segments.

Methods for Applying Cohesion Analysis

In the previous chapters, I established a theory of cohesion, continuity, and coherence. The following summarizes that theory.

Principles of Coherence (Holistic Judgments)

Principle of Continuity: Prefer a discourse in which the information appears relevant and connected.

Principle of Manner: Prefer a discourse in which the information is conveyed clearly.

Principle of Quantity: Prefer a discourse in which the amount of information appears adequate but not superfluous.

Principle of Quality: Prefer a discourse in which the information appears accurate or at least well supported.

Principles of Local Cohesion (Connections between Individual Discourse Elements)

Subprinciple of Similarity: Establish a sense of continuity where there are similar discourse elements.

Subprinciple of Proximity: Establish a sense of continuity where there are spatially or temporally proximate discourse elements.

Principles of Global Cohesion (Discontinuity Creating Discourse Organization)

Subprinciple of Intensity: Organize a discourse into segments based on the relative degree of the sense of continuity.

Subprinciple of Size and Symmetry: Organize a discourse into segments of adequate size and roughly symmetrical distribution.

Metaprinciples (Interaction of Coherence/Cohesion Principles)

Principle of Reinforcement: Strengthen the sense of coherence/continuity where principles support the same sense of coherence or continuity.

Principle of Conflict: Weaken the sense of coherence/continuity where principles support a different sense of coherence or continuity.

In this chapter, I explore how cohesion analysis should be applied in research. To this end, I first outline the methodology used to analyze cohesion in previous studies. Second, I address two previous complaints regarding cohesion analysis: the use of quantitative measures of cohesion and the exhaustive analysis of each cohesive element. And, third, I remind my readers of the justification used in chapter 1 for developing the theory and methodology presented here.

PREVIOUS APPLICATIONS OF COHESION ANALYSIS

Unfortunately, previous research has often noted dissatisfaction with the exhaustive analysis of cohesive elements in which every semantic cohesive element is marked and subjected to qualitative and quantitative analysis. That research has been based on Halliday and Hasan's (1976, pp. 329–339) methodology. In particular, for each successive sentence in the discourse, the analyst:

1. Lists the number of cohesive ties.
2. Lists each cohesive semantic element.
3. Notes the type of cohesion (i.e., reference, substitution/ellipsis, conjunction, or lexical) involved with each cohesive semantic element.
4. Notes the other element that is tied to the cohesive semantic element.
5. Notes the proximity of the cohesive element in this sentence and the other element in the tie.

To briefly illustrate the use of this methodology, let's consider a short segment of the oral discourse reproduced in Fig. 5.1. The results of the exhaustive cohesion analysis of semantic elements in sentences 5 and 6 of the discourse in Fig. 5.1 are given in Table 5.1. The results include: (a) the number of cohesive ties involved in each sentence (e.g., four in sentence 5); (b) a list of the cohesive semantic elements (e.g., *first* in sentence 5); (c) a categorization of the types of cohesion (e.g., *first* in sentence 5 is a type of conjunction); (d) a list of the elements to which the cohesive elements are tied (e.g., as a conjunction, *first* in sentence 5 creates a tie with all of sentence 4); and (e) a measure of the proximity of the elements involved in the tie (e.g., the tie between *first* and sentence 4 is immediate in that there are zero intervening sentences between the two elements).

Although it is entirely possible to simply expand the kind of coding techniques in Table 5.1 to include a wider range of discourse elements (e.g., visual) and a larger number of relationships among elements (e.g., intensity of similarity) as suggested by the theory developed in this book, my aim in this chapter is more ambitious. I want to demonstrate how the theory developed in this book can be most fruitfully applied to the analysis of discourse by addressing previous complaints about the application of cohesion analysis. To this end, I focus on two previous complaints about cohesion analysis. First, researchers have complained that quantitative

---

okay. today I'm gonna tell you a little bit about uh sêrvicing the main landing gear
[*very strong eye contact* . . . .
on a F fifteen E aircràft. first of all we need to determine whether the aircraft is on
. . . .]
jacks or on the ground. but in either case uh the sêrvicing is the same. or slightly
different. but uh, there's fôur procedures you always go through, uh before
[*four fingers held up*]
you sêrvice an aircràft.

{ first is you determine whether the aircraft is either normal
weight or heavyweight. and the differential is fifty-six hundred, or fifty-six
thousand pounds. thén uh you determine what the ambient air temperature is, and
you usually do that by calling up the the mach, and uh let them tell you what it is
over the radio. and thírd is uh you determine whether it's gonna need nitrogen or
hydraulic sèrvicing. and remêmber whenever you use nitrogen or or filtered dry air
sêrvicing you always use uh êye protection. okay, and then the fóurth thing is you,
[*hand to eye*]
uh and as i said before you determine whether the aircraft is on jacks or not.

um basically uh that's it.

FIG. 5.1. A transcribed account of an oral briefing demonstrating the use of
Halliday and Hasan's methodology for cohesion analysis.

measures of cohesion are of negligible use. To address this complaint, I explore the
relationship between cohesion and coherence. Second, researchers have complained about the amount of data produced in exhaustive analyses of cohesive
elements. To address this complaint, I provide a brief outline of four sample
applications of cohesion analysis.

TABLE 5.1
Results of Cohesion Analysis for Sentences 5 and 6 in the
Transcribed Discourse Shown in Fig. 5.1 Using the
Methodology Outlined in Halliday and Hasan (1976).

	# of Cohesive Ties	Cohesive items	Type of Cohesion	Presupposed Item	Distance
Sentence 5	4	first	conjunction: numerative	Sentence 4	immediate: preceding sentence
		first	lexical: hyponym	four	immediate: preceding sentence
		aircraft	lexical: same item	aircraft	immediate: preceding sentence
		heavyweight	lexical: hyponym	weight	immediate: same sentence
Sentence 6	3	and	conjunction: additive	Sentence 5	immediate: preceding sentence
		fifty-six	lexical: same item	fifty-six	immediate: same sentence
		thousand	lexical: overlap	hundred	immediate: same sentence

QUANTITATIVE MEASURES OF COHESION

Two of the measures associated with previous applications of cohesion analysis can be considered quantitative: (a) the total number of cohesive ties per sentence and (e) the number of intervening sentences between two cohesive elements in a tie. In terms of (e), if the analyst desires an analysis of the full range of discourse elements involved in cohesive ties, the appropriate unit of measurement is often not related to numbers of sentences. Specifically, the analyst of visual cohesive elements in a written document might gain more insight into the proximity of those elements by measuring in *inches* or *number of intervening pages* rather than by *number of intervening sentences*. Similarly, the analyst of auditory cohesive elements in a hypertext document might gain more insight into the proximity of those elements by measuring in *number of keystrokes, number of screens*, or in *seconds*. These are decisions which the analyst must confront based on his or her goals and the original goals of the producer of the discourse under analysis. I take up this topic again in the following section, but first I must address the quantitative measure that has met with the most serious criticism in the past.

In terms of (a), it is measures of *total number of cohesive items per sentence* that have earned cohesion analysis less than enthusiastic support. As Hartnett (1986) noted,

> Quantitative measures of cohesion have not seemed much more useful as an index of the quality of adult writing than has the incidence of other quantifiable features, such as [sentence] length, T-unit size, or mechanical and usage errors. (p. 143)

To illustrate this problem, consider Table 5.1 again. Specifically, we see that sentence 5 has four semantic cohesive ties, compared to only three ties in sentence 6. Unfortunately, the meaning of such quantitative differences in cohesion analysis is unclear: Is sentence 5 more effective than sentence 6? Would three semantic cohesive ties be sufficient to guarantee that sentence 6 is acceptable to recipients? Would ten semantic cohesive ties make sentence 6 unacceptable to recipients? The list of questions might continue ad nauseam. The point here is that quantitative analysis of semantic cohesive elements has met with little if any success in predicting coherence. In order to address the problem with quantitative measures of cohesion, we must consider in detail the relationship between cohesion and coherence in the remainder of this section.

The Interaction of Cohesion and Coherence

As Tierney and Mosenthal (1983) complained, the prevalence of semantic cohesive elements is so great in discourse that it "severely diminishes the usefulness of the cohesion concept as an index of coherence at the global or local level" (p. 228). This conclusion amounts to a claim that cohesion is not a sufficient, or perhaps even necessary, condition for guaranteeing coherence; this claim has in fact been widely recognized (e.g., Cooper, 1988; Giora, 1985; van Dijk, 1980; Witte & Faigley, 1981). As Bókay stated, it has "been impossible to find a finite set of rules, the theory of whose relations produce the coherence of these texts" (p. 414).

Halliday and Hasan (1976) provided the following example that can be used to illustrate this problem.

Example 5.1a.
Although the light was on he went to sleep. Although the house was unfurnished the rent was very high. Although he was paid a high salary he refused to stay in the job. (p. 19)

Note that the producer of this discourse has established some sense of continuity through the use of similar syntactic elements (i.e., each sentence is comprised of an introductory subordinate clause, followed by one main clause), similar morphological elements (i.e., each verb is in the simple past tense), similar semantic elements (i.e., the lexical items *although* and *was* occur in each introductory clause). In addition, the same sense of continuity is reinforced by the proximity of the three sentences (i.e., they are contiguous).

However, as Halliday and Hasan noted, a recipient's holistic sense is that Example 5.1 is not a coherent discourse. Interestingly, despite the presence of local cohesive elements that establish a sense of continuity, the incoherence of Example 5.1 can be attributed to the lack of continuity. There is certainly a lack of explicit continuity between the propositions expressed in each successive sentence, and it is nearly impossible to attain implicit continuity by building a bridge. Thus, we can note that the sense of continuity established through the local cohesion principles of similarity and proximity (as proposed in chapter 3) conflicts with the sense of holistic continuity described by the coherence principle of continuity (as proposed in chapter 2).

In fact, Halliday and Hasan used Example 5.1 to argue that non-semantic connections (e.g., syntactic parallelism in this case) in discourse are not sufficient to guarantee coherence, and that semantic connections are crucial. Unfortunately, according to the authors' theory, we must note that there is some semantic cohesion in Example 5.1, namely the repetition of lexical items, which the authors themselves classify as a kind of lexical cohesion. Therefore, apparently neither the presence of semantic nor non-semantic cohesive elements guarantees that a recipient will sense coherence.

The Precedence of Principles of Coherence

Hendricks (1988) provided a hint of the resolution to this problem: "[i]t does not make sense to have text analyses along the line of Halliday and Hasan's pursued independently of approaches which focus on global structure" (p. 105). Thus, in terms of the theory developed in this book, I maintain that cohesive elements increase discourse coherence only when they reinforce the same holistic sense of coherence as the coherence principles of continuity, manner, quantity, and quality. For instance, reconsider Example 5.1 within its original context. As Halliday and Hasan (1976) explained, "[Example 5.1] is in fact taken from a textbook of Chinese for English-speaking students.... It is the context of situation of this passage, the fact that it is part of a language textbook, that enables us to accept it as a text" (p. 20). Establishing the context of the discourse provides recipients with the knowledge they need in order to build a bridge that establishes a sense of holistic continuity in Example 5.1 and makes the discourse implicitly coherent. For example, the recipient who knows that this discourse appears in a second-language acquisition textbook can conclude that this discourse excerpt is designed to make a point about English sentences instead of about the propositions expressed by those sentences. Therefore, it is crucial to recall from chapter 2 that the recipient's holistic sense of coherence is always a function of his or her knowledge.

Stated another way, the producer of a coherent discourse has established the appropriate continuity, manner, quantity, and quality of information in the discourse explicitly (by providing recipients with the relevant knowledge) and implicitly (by assuming only that knowledge that recipients actually possess and can use to build bridges with). In the case of Example 5.1, when Halliday and Hasan used this excerpt without providing the readers of their book with knowledge of the topic or purpose of the discourse from which they took the example, the authors were withholding the very knowledge that their readers need in order to sense the continuity of the excerpt. Nevertheless, the use of cohesive elements, including semantic ones, which produce some sense of local continuity, does not guarantee the recipient's sense of the holistic continuity or coherence of a discourse.

The Relationship of Parts to the Whole

This inability to understand discourse coherence by establishing its "building blocks" (e.g., local cohesive elements) is analogous to the inability to understand visual or auditory "wholeness" as the sum of individual perceptions. As Gestalt psychologists made very clear, they believed that our perception of wholes is not a matter of simply adding up elements from "below" but of working from "above." As Wertheimer (1938) explained:

> Epistemologically this distinction between "above" and "below" is of great importance. The mind and the psychophysiological reception of stimuli do *not* respond after the manner of a mirror or photographic apparatus receiving individual "stimuli" *qua* individual units and working them up "from below" into the objects of experience. Instead response is made to articulation as a whole .... It follows that the apparatus of reception cannot be described as a piecewise sort of mechanism. It must be of such a nature as to be able *to grasp the inner necessity*

of articulated wholes. When we consider the problem in this light it becomes apparent that pieces are not even experienced as such but that apprehension itself is characteristically "from above." (p. 88)

More succinctly, Gestalt theory recognizes that the psychological whole is greater than the sum of its visual or auditory parts.

This notion seems especially appropriate to our application of Gestalt principles to explain discourse continuity established through local cohesive elements because we all recognize that the holistic sense of continuity in examples such as 5.1 cannot be characterized as simply the sum of all the cohesive relations between discourse elements described by the local cohesion principles of similarity and proximity or even by the global cohesion principles of intensity, and size and symmetry. In fact, holistic continuity cannot even be characterized by the metaprinciples of reinforcement and conflict if we consider only cohesive elements. Instead, we must recognize that the principles of coherence (i.e., continuity, manner, quantity, and quality) cannot be defined as the sum of local or global cohesive relations. Thus, it is not surprising that previous research that has used quantitative measures of semantic cohesion to predict coherence has met with little success.

In addition, because coherence is founded on the knowledge of a recipient and that knowledge is not only variable among recipients but partly idiosyncratic, we may never be able to predict coherence in the same way that modern linguistics can predict grammaticality. I am not ignoring the fact that computational linguists have been working on this task. But I am acknowledging that successes in this type of research have come in situations involving highly specific knowledge.

For example, Grishman (1990) reported a highly successful natural language-processing system that can parse even ill-formed (i.e., relatively incoherent) discourse. However, the discourses are equipment failure messages dealing with the starting air system for gas turbine propulsion systems aboard naval vessels. In other words, the knowledge required to build bridges that will make these reports coherent is highly specific, thus making it relatively easy to predict what knowledge is required to guarantee coherence. As Grishman (1990) noted:

> Our approach is based upon a very rich domain model. . . . [I]n the long term this will be a serious concern for both practical and theoretical reasons. . . . The theoretical concern is that we seem to have elaborated in the model more detailed information about the equipment than knowledgeable people require to understand these messages. (p. 55)

In brief, I cannot claim to have a theoretical model for predicting the specific knowledge that a recipient requires in order to sense coherence beyond the general principles introduced in chapter 2: that each recipient who finds what he or she considers appropriate continuity, manner, quantity, and quality of information in a discourse, based on his or her own knowledge, will find that discourse coherent.

Preference Versus Necessary and Sufficient Conditions

Having said this, I do not believe that we can ignore the fact that these local and global cohesive relations help us explain recipients' preferences regarding which discourses exhibit continuity and how discourses are organized. Gestalt psychologists made tremendous advances in our understanding of the human mind and visual/auditory perception despite the fact that the local perceptions (e.g., proximity) they noted did not "add up to" the holistic one (e.g., wholeness). Similarly, I believe that theories such as Halliday and Hasan's or the one presented in this book do advance our understanding of discourse despite the fact that local cohesive elements do not add up to holistic coherence or even holistic continuity. However, we cannot expect meaningful results from our applications of cohesion analysis if we continue attempts to correlate the total number of cohesive elements with recipients' holistic perceptions of coherence.

As Randquist (1985) reminded us, textual coherence is "a *relative* quality" (p. 192). Therefore, I know that, as a recipient, the cohesive elements in Example 5.1a caused me to see some continuity in that discourse and, therefore, made me more likely to continue searching for knowledge with which I could build a bridge that would make this discourse coherent for me. In contrast, I know that the lack of cohesive elements in Example 5.1b encourages less hope that continuity and coherence can be established:

Example 5.1b.
The light was on although he went to sleep. Despite the fact that the house was unfurnished the rent was very high. He was paid a high salary even though he refused to stay in the job.

This example certainly displays less continuity and coherence than the decontextualized Example 5.1a, although its only distinguishing feature is the lack of cohesion (i.e., no parallel syntax and no repetition of lexical items). As I have argued throughout this book, a recipient's sense of either holistic coherence or cohesion is clearly gradable or relative. My sense of continuity and coherence gradually increases as I first consider Example 5.1b, and then Example 5.1a, without any contextual knowledge, and finally Example 5.1a with knowledge of its original context.

If discourse coherence is a gradable and variable phenomenon, then the search for a set of necessary and sufficient conditions that insure coherence appears ill-conceived. Hatakeyama et al. (1985) made it clear that not all principles related to discourse must be conceived of as necessary and sufficient conditions: "[W]e regard it as a fundamental task of textological research to investigate, [sic] for which of these properties can intersubjectively acceptable, necessary and sufficient conditions be defined, and for which cannot [sic]" (p. 88). Instead, what we must search for is a set of conditions that describe the relative strength/ weakness of discourse coherence, no subset of which will guarantee such coherence, what Lerdahl and Jackendoff (1983) called "preference conditions." These authors note that, in musical "texts," there is no set of necessary and sufficient conditions that will predict the way in which listeners cognitively organize the structure of the piece; instead they propose a theory composed of conditions that express music "recipients' " preferences for structuring musical pieces. The coherence and cohesion principles set out in this book are similar to such conditions because they express the gradable and variable preferences of discourse recipients regarding their sense of the coherence of discourse.

It may be instructive at this point to consider another version of lines 2–11 of the memo reproduced in Fig. 4.4:

Example 5.2.

TO: Software Development Group
FROM: Suzanne Champion

DATE: April 8, 19—
SUBJECT: Trip Report
               Training for Automatic Instrumentation
               Corporation
               Glenview, IL
               April 3-5, 19—

This three-day course was an on-site introduction to our software. All 14 participants in my training group were new to computing, so the programming tasks focused on two areas. . . .

Note that Example 5.2 differs from the original in only two ways: (a) more discontinuity is established through the insertion of more white space between the FROM and DATE lines of the memo; and (b) less discontinuity is established through the deletion of some white space between the heading and body of the memo. These changes signal no changes in local cohesion but do mean that the principle of intensity, with which we justified the division of the discourse into two segments earlier, does not apply because the principle of proximity creates equal degrees of discontinuity in the heading (i.e., between FROM and DATE) and between the heading and body. Most importantly to my purpose here, note that I have continued to label the segments of the discourse in the same way despite this fact. In other words, although I find this version slightly less coherent than the original, my sense of the organization of the memo has not changed.

How can I account for this fact without negating the utility of cohesion analysis? First, as I stated earlier, it is clearly the case that holistic perceptions of coherence take precedence over local or even global cohesive elements. In Example 5.2, it is my socially constructed or culturally learned knowledge about the text type "memo" that prompts me to continue to see the same organization in the altered version as in the original despite the fact that there is no local or global cohesion cue that marks a distinctive discontinuity between the heading and the body.

Second, and most important, it is also clear to me that the original version is preferable precisely because the local cohesive elements and the global cohesion established through the principle of intensity reinforce the same sense of continuity as my holistic sense. Thus, I see sufficient motivation for arguing that cohesion enhances the relative degree of continuity, hence coherence, in discourse.

Document design research appears to support this conclusion. As Duin (1989) noted:

> On the one hand, a reader without appropriate prior knowledge about a passage's content cannot call up the appropriate schemata, and therefore cannot understand what is being read. On the other hand, if a reader does have appropriate prior knowledge, a text that is poorly organized will not elicit the appropriate schemata from the reader's mind. The organization of a text influences the way we acquire, remember and use information; it promotes efficient learning; it produces better recognition and recall; and it increases the likelihood of transfer of new knowledge to future tasks. (p. 98)

Thus, I propose that cohesion analysis provides an effective tool for diagnosing problems related to the relative strength of a recipient's sense of coherence. In turn, it seems promising to investigate the relationship between cohesion and the relative ease with which coherence is sensed by a recipient: in other terms, document *usability*. I address this suggestion briefly in the following section.

In sum, this section has addressed the relationship between cohesion and coherence in order to explain the basis for previous complaints about the use of quantitative measures of cohesion. I have acknowledged that the presence of cohesion does not guarantee coherence. I have shown that this problem is parallel to the Gestalt notion that a visual whole is greater than the sum of its parts. In addition, because we are not able to predict the specific knowledge of a recipient, I have proposed that coherence takes precedence over cohesion. Moreover, I have noted that cohesion principles must be viewed as conditions that describe recipients' preferences about coherence in discourse rather than as necessary and sufficient conditions that guarantee coherence. Finally, I have proposed that cohesion analysis might be better suited to the diagnosis of usability problems (i.e., the relative ease with which coherence is sensed) than to the prediction of coherence.

EXHAUSTIVE ANALYSES OF COHESIVE ELEMENTS

The desire for quantitative measures in previous applications of cohesion analysis has required that the analyst locate each and every semantic cohesive element in the discourse under investigation. The results of such exhaustive methods have produced another complaint about cohesion analysis exemplified by Hendrick's (1988) comments:

It takes Halliday and Hasan about seven pages to explain their scheme for coding the types of cohesion. . . . And when one imagines the whole text of, say, *Alice in Wonderland*, subjected to such an analysis, the result is bound to be a mass of data so overwhelming as to be practically useless. (p. 104)

As a concrete illustration, consider Table 5.1 again. Note the amount of data produced through an exhaustive analysis of just the semantic cohesive elements in the very brief, two-sentence segment of the discourse in Fig. 5.1. Moreover, because the theory of cohesion developed in this book accounts for the cohesive effect of the full range of discourse elements (e.g., visual as well as semantic) and for various types of relationships (e.g., intensity, reinforcement, etc.), the amount of data would increase significantly if we applied the same type of exhaustive analysis to the full range of cohesive elements and principles. More importantly, as I have argued in the previous section, past research makes it clear that cohesion analysis is not an appropriate method for predicting a recipient's holistic sense of coherence. Thus, the analyst who performs an exhaustive analysis with such goals will certainly be disappointed. It is not yet clear whether cohesion analysis can provide an appropriate method for predicting the ease with which a recipient senses continuity (i.e., one aspect of a document's usability).

In any case, I can only conclude that complaints about the amount of data analyzed through the application of cohesion analysis would cease if analysts found more of the analysis meaningful. Thus, I suspect that the problem actually concerns the ratio between the amount of data analyzed (because of the amount of effort required to analyze it) and the degree of insight gained via an analysis of that data. Happily, because quantitative measures of cohesion appear ill-advised, there is no longer any a priori requirement for the exhaustive analysis of cohesive elements. Instead, the amount of data analyzed should be determined by the scope of the questions of interest to the analyst.

Thus, there is no need to exhaustively analyze each cohesive element in a discourse unless the goals of the analysis require it. In many cases, a sufficiently focused research goal will limit the scope of cohesive elements that must be analyzed, thereby creating a more acceptable ratio between the amount of data analyzed (including the amount of effort required to analyze it) and the degree of insight gained via an analysis of that data. And, I believe, more tightly focused research goals will also enhance the quality of research in which cohesion analysis is an appropriate methodology. Unfortunately, past research has not always recognized that the quality of research applying cohesion analysis is dependent on the quality of the research questions under investigation. Nevertheless, Halliday and Hasan (1976) suggested as much:

In presenting a framework for the analysis and notation of a text, however, we should emphasize the fact that we regard the analysis of a text in terms of such a framework as a means to an end, not as an end in itself. There are numerous reasons why one might undertake such an analysis, and the enquiry will lead in all kinds of different directions . . . (p. 332)

In order to illustrate how the research goals of the analyst dictate the scope of cohesive elements under analysis, the remainder of this section briefly considers the potential application of cohesion analysis as a method of investigating various research topics. Because the range of potential topics that might be illuminated through the application of cohesion analysis is diverse, the discussion of the following topics is meant to be illustrative, not exhaustive.

Application 1: Cohesion and Usability

As I suggested earlier, cohesion analysis may be more useful for predicting document usability than coherence. Guillemette (1989) provided the following comprehensive definition:

> The term *usability* refers to the degree to which documentation [e.g., software manuals, bicycle assembly instructions, etc.] can be effectively used by target readers in the performance of tasks under environmental requirements and constraints. Effectiveness is defined in terms of reader performance with written materials and acceptability of those materials to the reader. Basic human performance dimensions include efficiency (speed) and bias (accuracy) in performing tasks. Readers themselves are the primary source for reporting perceptions of tiredness, discomfort, boredom, frustration, or excessive personal effort in using documentation. (p. 218)

Thus, I might hypothesize that cohesion produced through the use of visual similarity is more influential than that produced by syntactic similarity in determining usability (i.e., speed and accuracy with which tasks are performed).

The analyst interested in investigating this hypothesis might design a study in which his or her goal would be to determine the relative weight of syntactic versus visual similarity in predicting usability of a hypertext manual. Methodology would then be determined by the analysts' research goal. In this case, there would be no need to exhaustively analyze all cohesive elements in the hypertext manual. Instead, cohesion analysis/theory would be used in the design of various hypertext samples for testing speed and accuracy: Version A would contain both visual and syntactic cohesive elements; Version B visual but not syntactic; Version C syntactic but not visual; and Version D neither visual nor syntactic. Measures of the speed and accuracy with which a task was performed by recipients using each version of the manual could then be compared according to the cohesive elements present in each version. Those usability measures might be both objective and subjective: (a) accuracy could be objectively measured by noting the degree of success in completing an assigned task using the manual (e.g., 100% complete, 50% complete, etc.); and (b) accuracy could be subjectively measured by presenting recipients with a questionnaire containing questions eliciting their perceptions of the accuracy of the manual. (See Appendix B for samples of such questions based on the theory of coherence and cohesion presented here; current research is testing the validity and reliability of such questions for measuring coherence and usability.) Thus, the research goal of the analyst in this case would limit the amount of cohesive data with which he or she must deal.

However, the most important consequence of limiting the range of elements subjected to analysis is suggested by Mills and Dye (1985), who argued for the need for better design in usability research:

> In systematic research on usability, the effect of one factor at a time must be studied.... Systematic research ... can lead to principles for writing usable technical documentation. Having principles for writing usable documentation could reduce the massive rewrites that are sometimes required as a result of editing or testing. (p. 44)

I am arguing here that cohesion analysis can provide the foundation for one systematic approach to the study of the document design factors that appear to influence usability. And, consequently, such analyses might result in practical applications in the profession and pedagogy of document design.

Application 2: Cohesion, Usability, and SGML

The desire for the exchange and re-use of electronic information across all types of computer hardware and software has lead to the development of Standard Generalized Markup Language (SGML). The use of SGML for composing documents results in more standardization within document types. For instance, documents of the type Memo might be defined so that they must all include the element **header**, which would be composed of four subelements: **recipient**, **author**, **date**, and **subject**. In addition, **date** might be defined so that the only acceptable characters would be numerals (i.e., *January* could not be entered). Furthermore, **date** might be defined so that those numerals would always appear in italic type. (For those who are interested in learning about SGML, see issues of *Technical Communication* for the second and third quarters of 1993 or see Smith, 1992.)

The important point here is that the use of SGML standardizes much of the design of a particular type of document, including formatting. Standardization or consistency has often been noted as "the key to usable online information . . ." (Bradford, 1988, p. 204). Thus, I hypothesize that its use results in more cohesive, hence usable, documents. An analyst interested in exploring such a claim might design a study in which he or she seeks to establish a correlation between usability and cohesion in two software manuals: one created with SGML and one without. Again, methodology would be determined by the analysts' research goal. In this application, there would be no need to do an exhaustive analysis of every cohesive element in the two versions of the manual. Instead, the analyst might use cohesion analysis only where the two versions of the manual differed. As in Application 1, the use of cohesion analysis in Application 2 provides one systematic approach to the description of differences in the usability of versions of sample discourses.

Application 3: Cohesion and Organization in Oral Discourse

Despite its obvious importance for professional communicators, the effective design of oral communication receives relatively little attention. As an illustration, consider the fact that the ATTW's comprehensive bibliography of publications in technical communication for 1992 includes only two research articles and two pedagogical articles related to oral communication (Philbin, 1993). Thus, an analyst interested in descriptive research focusing on oral communication might contribute significantly to the current state of pedagogy in professional communication.

As one example within this realm, I hypothesize that the use of auditory cohesive elements reinforces perceptions of organization. An analyst interested in investigating this hypothesis might engage in a study in which his or her goal would be to establish a relationship between perceptions of the organization of a research paper presentation and the presence of auditory cohesion. This goal, in turn, would determine the appropriate methodology for applying cohesion analysis. In this case, an analyst might document possible organizations by introspection/expert review of the oral presentation and then note the presence or absence of auditory cohesion. Most important to my purpose here, note that the analyst need not code every cohesive element in the discourse under investigation. Instead, in this case, the analyst narrows the scope of elements analyzed to auditory elements.

Application 4: Variation in Oral, Written, and Electronic Discourse

Previous research investigating variation between oral and written language has been criticized because most studies have not carefully controlled for factors other than media type (e.g., Akinnaso, 1982, p. 110). In other words, researchers have compared the qualities of oral conversation with the qualities of written essays and then attempted to make conclusions about the nature of oral versus written language despite the rhetorical differences between conversations and essays (i.e., conversations tend to be informal, primarily social in function, etc. compared to essays, which tend to be formal and primarily persuasive in function).

Thus, to learn more about variation due to media, I hypothesize that visual elements are used differently in oral, written, and hypertext discourse. The analyst involved in testing this hypothesis might design a study in which a portion of a written instructional manual is used to develop an oral and hypertext version of the same instructions; the visual elements in the three media samples might then be compared. This goal determines the scope of elements that need to be analyzed.

In this case, all visual elements would need to be noted; thus analysis of visual cohesive elements would constitute only one aspect of the methodology.

In sum, previous researchers have complained about the amount of data that must be analyzed in the application of cohesion analysis based on Halliday and Hasan's methodology. I have framed this problem as dissatisfaction with the ratio between the amount of data analyzed and the degree of insight gained via that analysis. The four applications just covered were designed to illustrate how researchers might use cohesion analysis more fruitfully. By forming more tightly focused research questions, I have reduced the scope of data under investigation in each example application. In other words, in each of the four sample applications, I have suggested cohesion analysis as a methodology for investigating highly specific research questions. Therefore, I expect a more acceptable ratio between the amount of data analyzed and the degree of insight gained via that analysis in each of these applications.

CONCLUSIONS

The justification for developing this theory depended on two observations made in chapter 1. First, I noted a perceived need within the field of technical and scientific communication for research that provides general principles that form a foundation for evaluating document quality in pedagogical and professional practice by answering the following questions based on Shriver (1989a):

1. What are the various types of knowledge writers and readers bring to the task of communication production and comprehension?
2. What principles describe effective document design?
3. How can we develop effective methods of evaluating text quality?

Second, I noted that the answers to these questions were dependent on a theory of coherence and cohesion that had yet to be developed within linguistics or psychology. That theory needed to address the following questions:

A. How can we account for the unifying effect of the full range of discourse elements: semantic elements, other linguistic elements, and non-linguistic elements? What is the relationship between semantic elements and other cohesive elements?
B. What role do cohesive discourse elements play in establishing coherence? Can we predict when they will and will not enhance coherence?

Thus, I set out to develop a theory of discourse coherence and cohesion that would provide a framework within which the three, original questions could be approached. Based on this theory, I can now argue that the unifying effect of the full range of discourse elements is described by two cognitive perceptual principles that establish a sense of continuity: similarity and proximity, originally used in response to interaction with visual and auditory phenomena. In addition, I can argue that the cognitive perceptual principle of reinforcement explains how non-semantic cohesive elements can be used to reinforce semantic connections in discourse (Question A). Furthermore, this theory suggests that cohesion principles describe preference conditions and not necessary or sufficient conditions for establishing coherence; in addition, cohesion will enhance coherence only when it reinforces a recipient's holistic perception of coherence (Question B).

This theoretical framework, in turn, provides a way of approaching the more specialized concerns of technical and scientific communication professionals regarding effective document design. Based on this theoretical framework, discourse participants bring four types of knowledge to the tasks of discourse production and comprehension: linguistic (non-semantic) knowledge, pragmatic discourse knowledge, socially constructed (non-discourse) knowledge, and idiosyncratic knowledge. Moreover, this knowledge is gradable and variable, sometimes even unpredictable, among participants (Question 1).

Again, based on this theoretical framework, we can form a variety of hypotheses about document design: (a) cohesion produced through the use of visual similarity is more influential than that produced by syntactic similarity in determining usability (i.e., speed and accuracy with which tasks are performed); (b) use of SGML results in more cohesive, hence usable, documents; (c) the use of auditory cohesive elements reinforces perceptions of organization; and so on. These hypotheses might direct future investigations that seek to predict effectiveness of a particular design (Question 2). As I noted in chapter 1, my hope here is to spur more research in document design based on perceptual principles, not to provide a definitive theory that defines all practical guidelines for designing effective discourse.

Finally, this theoretical framework predicts some of the successful/unsuccessful components of methodologies designed to evaluate text quality:

1. recipient-focused evaluation methods (e.g., cloze testing, keystroke protocols, performance testing, and surveys [Shriver, 1989b, p. 242]) collect the only type of data that appropriately measure discourse coherence because a variety of recipients ordinarily interact with the discourse under investigation.

2. expert-judgment-focused evaluation methods (e.g., peer review, editorial review, and subject matter review [Shriver, 1989b, p. 242]) less appropriately measure coherence since only one recipient ordinarily interacts with the discourse under investigation.

3. text-focused evaluation methods (e.g., computerized stylistic software, guidelines, and checklists [Shriver, 1989b, p. 242]) least appropriately predict coherence because these methods involve no human discourse recipients.

Furthermore, because our theoretical framework describes the ways in which global cohesion creates perceptions of organization in discourse, it might be used to improve text-focused methods whose "inherent weakness ... lies in their predominant focus on word- and sentence-level features of the text" (Shriver, 1989b, p. 244). Despite researchers' warnings (e.g., Redish & Selzer, 1985) about the validity and reliability of such text-focused methods (especially readability formulas), they are seductive in terms of ease of use and cost. Thus, many researchers are interested in improving text-focused evaluation methods by developing quality metrics that go beyond traditional readability measures (Shriver, 1993, p. 245). The theoretical framework in this book might prove beneficial to developers of software whose aim is to evaluate document quality through computerized text analysis.

SUMMARY

In this chapter, I considered how the theory developed in this book might be applied in research. Based on previous complaints about cohesion analysis, I made two suggestions for future studies that use cohesion analysis. First, I argued against the use of quantitative measures of cohesion as an indicator of holistic perceptions of coherence. I noted that the presence of cohesion does not guarantee coherence, and thus cohesion principles are not necessary or sufficient conditions on coherence. Instead, I proposed that coherence takes precedence over cohesion and that cohesion principles are preference conditions that may be more strongly indicative of discourse usability than discourse coherence. Second, I argued against the a priori requirement for exhaustive analyses of cohesive elements. I presented four sample applications of cohesion analysis in research that reduced the scope of data under investigation because each pursued a tightly focused research question. Moreover, I argued that narrowly defined research goals might improve the quality of not only studies that incorporate cohesion analysis, but also of studies in document design in general.

CHAPTER 6

Implications for the Practice and Pedagogy of Technical and Scientific Communication

Although the primary focus of this book is on enhancing the quality of document design research, I feel strongly about the potential implications of the theory and methodology developed here for practice and pedagogy in technical and scientific communication. As I noted in chapter 1, communicators would like to be able to support their professional opinions by referencing published research (Brooks, 1991, p. 83). In addition, technical and scientific communication education relies on the ability to teach general principles that can be used to diagnose rather than simply detect communication problems (Flower et al., 1986, p. 47). The theory developed here is one means of providing general principles that form a foundation for professional education and practice. The need for systematic, general principles in education is nicely illustrated by Riley's hypothetical example in which a teacher must respond to an English as a second language (ESL) student who writes, *I will taking physics next semester.* As Riley (1988b) explained:

> The teacher who suggests a revision like "I will take" or "I will be taking" is providing accurate information, but only about this particular sentence. What the student needs is a more general principle that is both accurate and revealing: for example, "In an active sentence, the verb form following a modal (e.g., *will*) is always uninflected (e.g., *take*); the verb form following auxiliary *be* is always a present participle (e.g., *taking*)." (pp. 1–2)

Systematic, general principles are important in order to develop effective text-evaluation methods, whether those methods are used by teachers, professional communicators, or researchers. As one actual example, consider the discourse excerpt that follows. This example consists of an original, student-written excerpt from a set of instructions and one of the evaluative comments made by the student's technical communication teacher (denoted by the *script* font):

Example 6.1.

Do *not* use:

- baby powder (it is harmful to your baby's lungs)
- baby oils or lotions (they cause clogging of your baby's skin)
  ↑ *the parenthetical*
  explanations here
  work well

(Roberts, 1992, p. 218)

Note that the evaluator has offered what appears to be accurate feedback. In other words, the student writer's choice to use an orthographic cue, parentheses, to separate the information regarding materials (i.e., *baby powder* and *baby oils or lotions*) from the explanatory information about those materials is probably an effective one. Unfortunately, this feedback offers no revealing, general principle describing why or when parenthetical explanations are effective, thus lessening the possibility that the feedback can be applied to future writing tasks. (It is of course possible, even likely, that the evaluator has already discussed more general principles regarding document design or that the evaluator's comment is meant to encourage rather than inform the writer.) My point here is that knowledge of such general principles is what determines whether the writer understands what he or she has done well and should continue doing.

At a minimum, what a writer needs in such a situation is knowledge that (a) in order to increase the usability of instructions, lists of material and of tasks should be clearly separated from explanatory information and that (b) orthographic devices (e.g., parentheses or dashes) can be used to separate ideas in a written text. At best, what the writer needs is even more general knowledge about how unity (and discontinuity) is created in written texts by the full range of discourse elements so that he or she can apply this knowledge to plan the design of future documents or to diagnose their flaws. The theory of cohesion developed in this book provides one theoretical framework within which this general knowledge is described. Thus, professional writers who know the theory are in a position to use it to justify their professional opinions about design. Likewise, technical communication teachers who know the theory are in a position to use it to both justify their professional opinions about design to their students and to develop classroom activities that will teach this general knowledge to their students.

However, there have been significant misunderstandings in the past about the relationship between linguists (and their theories) and communication specialists (and their professional and pedagogical concerns). Thus, I want to turn my attention to this relationship in the next section of this chapter before I suggest some of the specific implications of my theory for document design practice and pedagogy.

THEORY, APPLICATION, AND PRACTICE

Although it is commonplace to treat application and practice as synonyms, as in "theory and application" or "theory and practice," Riley (1987) developed a theory that distinguishes between them.[1] The main problem with the two-pronged division, according to Riley, is that theory is typically too abstract, remote, and complex to be put directly into any sort of everyday practice. This is espcially true because the theory generally comes from one field (e.g., linguistics) and the practice is done in another (e.g., technical communication). Instead, Riley proposed a three-way dinstinction that keeps application and practice as separate entities. These three levels of activity can be differentiated according to two parameters: (a) whether the domain of inquiry is universal or restricted to a specific population, and (b) whether the purpose of the inquiry is to describe or prescribe.

The distinction between theory, on the one hand, and application and practice, on the other, is that **theory** is concerned with a universal account of a particular range of phenomena and **application** and **practice** are concerned with the relevance of theory to a specific sample population. Consider an example from chemistry. A free radical is an atom or compound in which there is an unpaired electron. The theory of free radicals is universal in the sense that it describes the behavior of free radicals regardless of the circumstances under which they occur (e.g., the specific compounds involved, the geography of where such compounds occur, etc.). This theory, however, has been applied to a specific sample population of chemical compounds, animal tissue, and ultimately has been used to develop a new surgical technique designed to minimize tissue damage and promote tissue healing. Now consider an example from linguistics. An indirect speech act is a speech act performed indirectly by performing another (e.g., directing someone to stand up by asking, "Can you stand up?" rather than telling him "Stand up"). The theory of indirect speech acts is universal in the sense that it describes the properties of such acts regardless of the circumstances under which they occur (e.g., who utters the act, what language the speaker is using, etc.). However, this theory has been applied to a specific sample population, writers of professional letters, and ultimately has been used to develop classroom exercises for improving professionals' control of tone.

In terms of this book, I developed a theory of coherence and cohesion in chapters 2–4. This theory is universal in the sense that it describes coherence and cohesion produced through the full complement of discourse elements (e.g., visual, morphosyntactic, semantic, etc.) in discourse conveyed through any medium (e.g., oral, written, etc.) regardless of the producer or recipient (e.g., first-grade student, medical researcher, Air Force officer, etc.), and regardless of the purpose or genre of that discourse (e.g., environmental impact statement, novel, conversation, etc.).

---

[1]This section is largely based on Parker & Campbell (1993).

However, in chapter 5 I turned from theory to application by suggesting how this theory might be applied in research studies dealing with specific sample populations (e.g., visual vs. syntactic similarity in a hypertext manual; auditory cohesive elements in an oral research presentation; etc.).

In contrast, the distinction between theory and application, on the one hand, and practice, on the other, is that **theory** and **application** are descriptive, whereas **practice** is prescriptive. Description simply provides an account of a range of phenomena; prescription, however, attempts to effect some sort of material change in the world. Consider our chemistry example once again. The theory of free radicals and its application to animal tissue are both descriptive in that the former describes a theory and the latter describes how the theory might be applied to one specific sample population; neither, however, actually effects change. On the other hand, the act of a surgeon putting the application into practice by using a particular technique that minimizes the creation of free radicals in animal tissue during surgery on a particular horse is prescriptive in that it is an attempt to change the chemistry of the horse's tissue. Now consider our linguistics example once more. The theory of indirect speech acts and its application to professional letter writing are both descriptive in that the former describes a theory and the latter describes how the theory might be applied to a specific sample population. In contrast, the act of a college instructor putting the application into practice, by conducting classroom exercises that require the analysis of direct and indirect speech acts, is prescriptive in that it is an attempt to change the behavior of those students. The distinctions among theory, application, and practice are summarized in Table 6.1.

In terms of this book, both the theory of discourse coherence and cohesion and its potential applications in a research study are descriptive, in that the former describes, for example, a theory of auditory cohesion and the latter describes the use of auditory cohesion and its effects on perceptions of organization in a research paper presentation. In contrast, the act of a scientific communication instructor putting that application into practice, by lecturing about auditory techniques students should use to create organization in effective oral presentations or by playing tapes of oral presentations and having students practice listening for auditory cues to the organization of different presentations, is prescriptive. In these cases, the goal of the instructor is to actually change the behavior of his or her students when they give oral presentations.

Thus, it is important to understand that my primary goal in this book has been theoretical: describing universal discourse phenomena. Nevertheless, I have sought to indicate how that description might be applied to research investigating effectiveness in document design in chapter 5. Although I mention some of the implications for professional practice and teaching that seem most apparent to me in the following section, precisely how these descriptive principles might be put into practice is beyond the scope of this book. Hopefully, I have presented this theory as clearly and explicitly as necessary in order to facilitate its use by any practitioner interested in designing specific pedagogical or design techniques.

TABLE 6.1
Summary of the Distinctions Among Theory, Application, and Practice

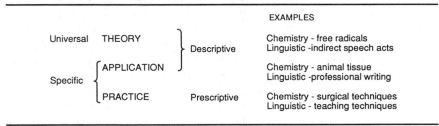

IMPLICATIONS OF THIS THEORY FOR DOCUMENT DESIGN

Based on my theoretical framework, I can make a number of predictions about effective document design.[2] However, it is important to note that these predictions are preliminary in that few are based on extensive results from previous research. Let me reiterate: My fundamental aim in this book is to provide a springboard for more research in document design based on perceptual principles. In the remainder of this chapter, I discuss the implications of this theory of coherence and cohesion in terms of the importance of (a) technical vocabulary, (b) audience analysis, (c) educating the audience, (d) consistency, and (e) layout. (See chapter 5 for a summary of the theory.)

The Importance of a Technical Vocabulary

My theoretical framework, though preliminary, can provide an explicit, descriptive vocabulary for discussing certain aspects of document design. Thus, a professional editor or teacher might use the technical terminology *cohesion based on similarity* as a way of defining a design problem rather than simply noting that a document "lacks unity," "is incoherent," or "doesn't hang together." Widespread adoption of such meaningful, technical terms might do much to encourage respect for the professionalism of technical and scientific communicators and their instructors. More often than not, these communicators are respected for their talent or skill in designing documents but not for their knowledge about design.

Many aspects of the theory might provide instructors with technical vocabulary for concepts important in document design. As one example, the principles of similarity, proximity, reinforcement, and conflict define the relationship between a document's meaning and its design. In other words, similar, proximate design

---

[2]Let me note that, although I follow convention in the field of technical and scientific communication by using the term *document* design here, my focus is not exclusively on written documents.

elements should be used to reinforce information with similar meanings in a document; on the other hand, dissimilar, nonproximate design elements should be used to signal information with dissimilar meanings. It is important to understand that, although the choice not to spend classroom time teaching the entire theory is certainly justifiable, this choice does not absolve teachers or other professionals from knowing such theories and basing their practice on them rather than on folklore or other untested advice. Fortunately, such theories can be used selectively in order to address the specific problems of one document or of one student. In sum, this theory of coherence and cohesion can provide professional communicators and teachers of technical and scientific communication with a technical vocabulary for discussing some of the concepts related to document design.

The Importance of Audience Analysis

The principles of coherence predict that effectively designed documents will have four qualities: relevance, clarity, adequacy, and accuracy. More importantly, this theory makes clear that these qualities are the result of the recipient's judgment about the document. Thus, it becomes clear why professional communicators and their instructors pay so much attention to the need for audience analysis: a producer cannot hope to design an effective document unless he or she knows what information his or her recipients think is relevant, clear, adequate, and accurate. Producers of documents must consider their recipients' knowledge of language, of discourse conventions or genres, and of the subject matter and the world. Clearly, these considerations are highly demanding.

Moreover, there is always the possibility that, despite valiant efforts by the producer, recipients may determine a document to be irrelevant, unclear, inadequate, or inaccurate based on some idiosyncratic knowledge that is, by definition, unpredictable. Thus, even when a professional communicator uses a good questionnaire to learn more about the audience for the software documentation he or she must write or even when an environmental engineering student plans meetings with an advisor to learn more about expectations for his or her dissertation, the potential for incoherence remains. In other words, producers should use any technique at their disposal for analyzing a document's audience, but recognize there are no guarantees. In brief, this theory suggests that audience analysis is a crucial aspect of effective document design.

The Importance of Educating the Audience

As I noted earlier, the audience's knowledge of language, genre conventions, and subject matter influence their judgments about the coherence of a document and its design. On the one hand, this fact dictates that designers of documents must educate their audience about these aspects if there is a possibility that the audience lacks such knowledge. For instance, it is necessary to include a glossary of terms in a physics textbook because some words in the document will be unknown to the audience. On the other hand, this fact mandates that educational curricula must educate the audience of documents about language, genre conventions, and subject matter. There is little argument about expanding students' knowledge of a subject matter within education. Less widely accepted, though, is the importance of expanding all students' knowledge of language and genre conventions. At all levels of education, most curricula designed to educate students about language focus on literature. Although I believe that literature is important, we have seen that it is the least applicable genre (in terms of design) for the majority of documents that students will receive and design in their professional future. In addition, expanding students' knowledge about language might better be accomplished by courses in linguistics, which explicitly teach the analysis of language. Furthermore, the primary focus of literacy education concerns written language, even though oral communication is crucial to professional success, and literacy in electronic communication is nearly as crucial now and will probably supplant written language as the most crucial in the future.

At a college level, freshman composition, as typically taught, focuses on the academic essay, again a peripheral genre. Fortunately, many professional curricula include speech courses and technical or scientific communication courses in which students are exposed to a wider and more relevant variety of genres and media. However, in my experience relatively little emphasis is given in these courses to reading or listening. The advent of programs in writing across the curriculum ensures that some students are being educated in a wider variety of discourse genres and perhaps media, but again the explicit emphasis is on written language and on the students' role as producer rather than recipient. In sum, the theory highlights the importance of educating the audience of documents in two ways. First, I have shown that writers and speakers must sometimes take the role of teaching their readers and listeners. Second, I have argued that educational curricula must expose students to a variety of genres and media of communication so that they have the prerequisite knowledge to comprehend the documents they are likely to receive.

The Importance of Consistency

The principle of similarity predicts that the coherence of a document is enhanced when design elements are used consistently. This general principle of design is widely noted by researchers (e.g., Ramey, 1988, p. 156) and teachers (e.g., Houp & Pearsall, 1992, p. 224) and appears to apply to many different aspects of design. For instance, we can predict that a book in which the beginning of each of its six chapters begins with a quotation is preferable to one in which just the second and fifth chapters do. As another example, we can predict that a briefing in which the same term is used throughout is preferable to one in which a variety of synonyms is used. In addition, we can predict that a hypertext software manual in which the entries on the table of contents screen match the headings on other related screens is preferable to one that does not. Finally, we can predict that a research article in which all of (and only) the technical terms appear in boldface is preferable to one in which only some of them do or in which boldface is also used for emphasis. In all of these cases, this theory of cohesion contains an explicit account of the need for such consistency: The principle of similarity establishes continuity and foregrounds dissimilarities between elements in a document.

The Importance of Page Layout, Screen Design, and Oral Arrangement

The principle of proximity predicts that the coherence of a document is enhanced when related pieces of information are designed near each other. Thus, for example, we can predict that assembly instructions for a bicycle that include a diagram of the frame that occurs next to a discussion of the parts of the frame is preferable to a set of instructions in which the diagram occurs on a different page. In addition, we can predict that an electronic book in which all related screens of information can be accessed consecutively is preferable to one that requires navigation through some screens of unrelated information. As a final example, we can predict that an Air Force briefing in which a hand gesture symbolizing the number 4 occurs simultaneously with the pronunciation of the word *four* is preferable to one in which the hand gesture occurs 5 minutes later. In brief, the importance of these design considerations is described by the theory of cohesion: the principle of proximity establishes continuity and foregrounds temporal and spatial distance between elements in a document.

SUMMARY

In this final chapter, I restated the observation that the practice and pedagogy of technical and scientific communication seeks general principles that can be used to justify professional opinion or to educate novices. In addition, I proposed a three-way distinction between theory, application, and practice in order to clarify the relationship between the theory developed in this book and the practical concerns of document design. Thus, I concluded by stating some preliminary implications of the theory of coherence and cohesion for the practice and pedagogy of document design, including (a) the availability of a technical vocabulary that would enhance the professionalism of our practice, (b) the importance of audience analysis in the design of coherent documents, (c) the need for broader educational curricula that expose students to a variety of genres and media for communicating, (d) the importance of consistency for producing continuity in the design of coherent documents, and (e) the influence of page layout, screen design, and oral arrangement for producing continuity in the design of coherent documents.

Appendix A:
Sources of Discourse Data

Adams, O. R. (1970). Stomach tube passage in the equine: Technique and uses. *Veterinary Scope, 15*(1), 1–7.

Benson, P. J., & Burnett, R. E. (1992). The shaping of written communication. In B. R. Sims (Ed.), *Studies in technical communication: Selected papers from the 1991 CCCC and NCTE meetings* (pp. 87–102). Denton, TX: University of North Texas Press.

Burnett, R. E. (1990). *Technical communication* (2nd ed.). Belmont, CA: Wadsworth.

Cather, W. (1987). Neighbour Rosicky. In D. McQuade (Ed.), *The Harper American literature* (Vol. 2, pp. 1043–1064). New York: Harper & Row.

Chaucer, G. (1987). The miller's tale. In L. D. Benson (Ed.), *The Riverside Chaucer* (3rd ed., pp. 68–77). Boston: Houghton Mifflin.

Chomsky, N. (1986). *Barriers*. Cambridge, MA: MIT Press.

Department of the Air Force. (1990). Technical orders for F-15E aircraft (hypertext version). Unpublished.

Department of the Air Force. (1991, March). Proposed conversion to F-16 C/D Squadron Eielson AFB AK. (Environmental assessment).

Faulkner, W. (1936). *Absolom! Absolom!* New York: The Modern Library, Random House.

Freestone, J. F., Shoemaker, S., & McClure, J. J. (1990). *Pulmonary abscessation, liver disease and IgM deficiency associated with* Rhodocuccus equi *in a foal*. Unpublished manuscript, School of Veterinary Medicine, Louisiana State University, Baton Rouge, LA.

Joyce, J. (1946). Eveline. In H. Levin (Ed.), *The portable James Joyce* (pp. 46–51). New York: Viking Press.

King, M. L., Jr. (1989). I have a dream. In D. Tannen (Ed.), *Talking voices: Repetition, dialogue, and imagery in conversational discourse* (p. 83). Cambridge, UK: Cambridge University Press.

Lear, E. (1986). Limerick. In M. H. Abrams (Ed.), *The Norton anthology of English literature* (5th ed., Vol. 2, p. 1591). New York: W. W. Norton.

Microsoft Corporation. (1987). *MS-DOS: User's guide and user's reference*. Author.

Paccamonti, D., Chang, S. T., Drost, M., Wilcox, C., Prichard, D., & Fields, M. J. (1990). *Induction of parturition in cows using porcine relaxin*. Unpublished manuscript, School of Veterinary Medicine, Louisiana State University, Baton Rouge, LA.

Parker, F. (1980). The perceptual basis of consonant cluster reduction and final devoicing. *Journal of Phonetics, 8*, 259–268.

Roberts, D. D. (1992). Commentary. In S. Dragga (Ed.), *Technical writing: Student samples and teacher responses* (pp. 207–236). Association of Teachers of Technical Writing.

Schafer, R. L., Evans, D. E., & Johnson, C. E. (1990). Some similitude aspects of multiple chisel systems. *Transactions of the ASAE, 33*(3), 777–782.

SCS Engineers. (1991). *Proposal to prepare design plans for closure of "X" canyon sanitary landfill.* Unpublished. Long Beach, California. Author.

Seuss, Dr. (1960). *Green eggs and ham.* New York: Beginner Books.

Swillum, J., Captain, USAF. (Speaker). (1992, December). Technical orders for F-15E aircraft. Unpublished briefing, Wright-Patterson Air Force Base, OH.

Tannen, D. (1989). *Talking voices: Repetition, dialogue, and imagery in conversational discourse.* Cambridge, UK: Cambridge University Press.

Tolkien, J. R. R. (1966). *The hobbit* (rev. ed.). New York: Ballantine Books.

U. S. Army Corps of Engineers. (1990). *Great Egg Harbor Inlet and Peck Beach, New Jersey Project.* (Final Supplemental Environmental Impact Statement).

van Ee, R., Gibson, K., & Roberts, E. D. (1990). *Osteochondritis dissecans in the lateral ridge of the talus in a dog.* Unpublished manuscript, School of Veterinary Medicine, Louisiana State University, Baton Rouge, LA.

Walsh, T., & Parker, F. (1983). The duration of morphemic and non-morphemic /s/ in English. *Journal of Phonetics, 11*, 201–206.

Appendix B

Sample questions for use in a survey designed to subjectively measure recipient perceptions related to coherence and usability.

COHERENCE
> I understood this _____.
> The author made this _____ understandable.

USABILITY
> I easily answered the questions I was given by using this _____.
> The author made this _____ easy to use in answering the questions I was given.

CONTINUITY
> The author made this _____ unified.
> All of the information in this _____ was related.
> The author made sure all of the information in this _____ was related.

CLARITY
> This _____ was clear.
> The purpose of this _____ was clear.
> The meaning of this _____ was clear.
> All of the information in this _____ was clear.
> The author made this _____ clear.

COMPLETENESS
> This _____ gave too little information
> This _____ gave too much information.
> The amount of information in this _____ was adequate.
> The author made this _____ complete.

ACCURACY
> This _____ was accurate.
> This _____ was well supported.
> The information in this _____ was true.
> The author made this _____ accurate.

References

Akinnaso, F. N. (1982). On the differences between spoken and written language. *Language and Speech, 25*(2), 97–125.

Anderson, P. V. (1987). *Technical writing: A reader-centered approach.* San Diego, CA: Harcourt Brace.

Anderson, R. C., & Pearson, P. D. (1984). A schema-theoretic view of basic processes in reading comprehension. In P. D. Pearson (Ed.), *Handbook of reading research* (pp. 255–291). New York: Longman.

Austin, J. L. (1962). *How to do things with words.* Oxford, UK: Clarendon.

Bamberg, B. (1983). Assessing coherence: A reanalysis of essays written for the national assessment of educational progress, 1967–1977. *Research in the Teaching of English, 18*(3), 305–319.

Barton, B. F., & Barton, M. S. (1985). Toward a rhetoric of visuals for the computer era. *The Technical Writing Teacher, 12*(2), 126–145.

Benson, P. J. (1985). Writing visually: Design considerations in technical publications. *Technical Communication, 32*(4), 35–39.

Benson, P. J., & Burnett, R. E. (1992). The shaping of written communication. In B. R. Sims (Ed.), *Studies in technical communication: Selected papers from the 1991 CCCC and NCTE meetings* (pp. 87–102). Denton, TX: University of North Texas Press.

Bernhardt, S. A. (1986). Seeing the text. *College Composition and Communication, 37*(1), 66–78.

Besner, D., & Humphreys, G. W. (Eds.). (1991). *Basic processes in reading: Visual word recognition.* Hillsdale, NJ: Lawrence Erlbaum Associates.

Bókay, A. (1985). Text and coherence in a psychoanalytic theory of jokes. In E. Sözer (Ed.), *Text connexity, text coherence: Aspects, methods, results* (pp. 414–438). Hamburg: Buske.

Bradford, A. N. (1988). A planning process for online information. In S. Doheny-Farina (Ed.), *Effective documentation: What we have learned from research* (pp. 185–211). Cambridge, MA: MIT Press.

Britton, B. K., & Black, J. B. (1985). Understanding expository text: From structure to process and world knowledge. In B. K. Britton & J. B. Black (Eds.), *Understanding expository text: A theoretical and practical handbook for analyzing explanatory text* (pp. 1–9). Hillsdale, NJ: Lawrence Erlbaum Associates.

Brooks, T. (1991). Career development: Filling the usability gap. *Technical Communication, 38*(2), 180–184.

Campbell, K. S. (1990). Explanations in negative messages: More insights from speech act theory. *Journal of Business Communication, 27*(4), 357–375.

Campbell, K. S. (1991). Structural cohesion in technical texts. *Journal of Technical Writing and Communication, 21*(3), 221–237.

Campbell, K. S. (1992). Repetition and cohesion in technical graphics. In B. R. Sims (Ed.), *Studies in technical communication: Selected papers from the 1991 CCCC and NCTE meetings* (pp. 73–85). Denton, TX: University of North Texas Press.

Charniak, E. (1972). *Toward a model of children's story comprehension* (TR-266). Cambridge, MA: MIT Artificial Intelligence Lab.

Charolles, M., & Ehrlich, M.-F. (1991). Aspects of textual continuity: Linguistic approaches. In G. Denhiere & J.-P. Rossi (Eds.), *Text and text processing* (pp. 251–267). Amsterdam: Elsevier.

Chomsky, N. (1976). Conditions on rules of grammar. *Linguistic Analysis, 2*(4), 303–351.

Clark, H. H., & Haviland, S. (1977). Comprehension and the given-new contract. In R. O. Freedle (Ed.), *Discourse production and comprehension* (pp. 1–40). Norwood, NJ: Ablex.

Cooper, A. (1988). Given-new: Enhancing coherence through cohesiveness. *Written Communication, 5*, 352–367.

Cooper, M. M. (1982). Context and vehicle: Implicatures in writing. In M. Nystrand (Ed.), *What writers know: The language, process and structure of written discourse* (pp. 105–128). New York: Academic Press.

Daneš, F. (Ed.). (1974). *Papers on functional sentence perspective, Janua Linguarum* (Suppl. Series Minor, No. 147). Mouton: The Hague.

Davidson, J. (1984). Subsequent versions of invitations, offers, requests, and proposals dealing with potential or actual rejection. In J. M. Atkinson & J. Heritage (Eds.), *Structures of social action: Studies in conversation analysis* (pp. 102–128). Cambridge, UK: Cambridge University Press.

de Beaugrande, R.-A., & Dressler, W. U. (1981). Introduction to TextLinguistics. London: Longman.

Derrida, J., & Olson, G. A. (1991). Jacques Derrida on rhetoric and composition: A conversation. In G. A. Olson & I. Gale (Eds.), *(Inter)views: Cross-disciplinary perspectives on rhetoric and literacy* (pp. 121–141). Carbondale, IL: Southern Illinois University Press.

Dorfmüller-Karpusa, K., & Dorfmüller, T. (1985). Some interdisciplinary remarks about coherence. In E. Sözer (Ed.), *Text connexity, text coherence: Aspects, methods, results* (pp. 555–566). Hamburg: Buske.

Duin, A. H. (1989). Factors that influence how readers learn from text: Guidelines for structuring technical documents. *Technical Communication, 36*(2), 97–101.

Dukes, T. (1988). I was a victim of the process approach. *Technical Writing Teacher, 15*(1), 78–83.

Fahnestock, J. (1983). Semantic and lexical coherence. *College Composition and Communication, 44*(4), 400–416.

Flower, L., Hayes, J. R., Carey, L., Schriver, K., & Stratman, J. (1986). Detection, diagnosis, and the strategies of revision. *College Composition and Communication, 37*(1), 16–55.

Giora, R. (1985). What's a coherent text? In E. Sözer (Ed.), *Text connexity, text coherence: Aspects, methods, results* (pp. 16–35). Hamburg: Buske.

Grice, H. P. (1975). Logic and conversation. In P. Cole & J. Morgan (Eds.), *Syntax and semantics* (Vol. 3, pp. 41–58). New York: Academic Press.

Grishman, R. (1990). Domain modeling for language analysis. In U. Schmitz, R. Schütz, & A. Kunz (Eds.), *Linguistic approaches to artificial intelligence* (pp. 41–58). Frankfurt: Verlag Peter Lang.

Grosz, B. J. (1977). *The representation and use of focus in dialogue understanding* (Tech. note 151). Menlo Park, CA: Stanford Research Institute Artificial Intelligence Center.

Guillemette, R. A. (1989). Usability in computer documentation design: Conceptual and methodological considerations. *IEEE Transactions on Professional Communication, 32*(4), 217–229.

Halliday, M. A. K., & Hasan, R. (1976). *Cohesion in English*. London: Longman.

Hatakeyama, K., Petöfi, J. S., & Sözer, E. (1985). Text, connexity, cohesion, coherence. In E. Sözer (Ed.), *Text connexity, text coherence: Aspects, methods, results* (pp. 36–105). Hamburg: Buske.

Hartnett, C. G. (1986). Static and dynamic cohesion: Signals of thinking in writing. In B. Couture (Ed.), *Functional approaches to writing* (pp. 142–153). London, UK: Frances Pinter.

Hendricks, W. O. (1988). Discourse analysis as a semiotic endeavor. *Semiotica, 72,* 97–124.

Henle, M. (1961). Some effects of motivational processes on cognition. In M. Henle (Ed.), *Documents of gestalt psychology* (pp. 172–186). Berkeley: University of California Press.

Houp, K. W., & Pearsall, T. E. (1992). *Reporting technical information* (7th ed.). New York: Macmillan.

Jacobson, R., & Pomorska, K. (1983). *Dialogues.* Cambridge, MA: MIT Press.

Johns, A. M. (1980). Cohesion in written business discourse: Some contrasts. *The ESP Journal, 1*(1), 35–44.

Kieras, D. E. (1985). Thematic processes in the comprehension of technical prose. In B. K. Britton & J. B. Black (Eds.), *Understanding expository text: A theoretical and practical handbook for analyzing explanatory text* (pp. 89–107). Hillsdale, NJ: Lawrence Erlbaum Associates.

Kinneavy, J. (1987). The process of writing: A philosophical base in hermeneutics. *Journal of Advanced Composition, 7,* 1–9.

Kintsch, W., & van Dijk, T. A. (1978). Toward a model of text comprehension and production. *Psychological Review, 85*(5), 363–394.

Lerdahl, F., & Jackendoff, R. (1983). *A generative theory of tonal music.* Cambridge, MA: MIT Press.

Lovejoy, K. B. (1987). The Gricean model: A revising rubric. *Journal of Teaching Writing, 6*(1), 9–17.

Manning, A. D. (1988). Literary vs. technical writing: Substitutes vs. standards for reality. *Journal of Technical Writing and Communication, 18*(3), 241–262.

Markels, R. B. (1983). Cohesion paradigms in paragraphs. *College English, 45*(5), 450–464.

Mathes, J. C., & Stevenson, D. W. (1991). *Designing technical reports* (2nd ed.). New York: Macmillan.

Mayer, R. E. (1985). Structural analysis of science prose: Can we increase problem-solving performance. In B. K. Britton & J. B. Black (Eds.), *Understanding expository text: A theoretical and practical handbook for analyzing explanatory text* (pp. 65–87). Hillsdale, NJ: Lawrence Erlbaum Associates.

McCulley, G. A. (1985). Writing quality, coherence, and cohesion. *Research in the Teaching of English, 19*(3), 269–282.

Miller, G. A., & Johnson-Laird, P. N. (1976). *Language and perception.* Cambridge, MA: Harvard University Press.

Mills, C. B., & Dye, K. L. (1985). Usability testing: User reviews. *Technical Communication, 32*(4), 40–44.

Moore, P., & Fitz, C. (1993). Using Gestalt theory to teach document design and graphics. *Technical Communication Quarterly, 2*(4), 389–410.

Myers, G. (1991). Lexical cohesion and specialized knowledge in science and popular science texts. *Discourse Processes, 14,* 1–26.

Parker, F., & Campbell, K. S. (1993). Linguistics and writing: A reassessment. *College Composition and Communication, 44*(3), 295–314.

Parker, F., & Riley, K. (1994). *Linguistics for Non-linguists* (2nd ed.). Boston: Allyn & Bacon.

Philbin, A. I. (Ed.). (1993). 1992 ATTW bibliography. *Technical Communication Quarterly, 2*(3), 447–475.

Ramey, J. (1988). How people *use* computer documentation: Implications for book design. In S. Doheny-Farina (Ed.), *Effective documentation: What we have learned from research* (pp. 143–158). Cambridge, MA: MIT.

Randquist, M. G. (1985). The barely visible glue: Some aspects of textual connectedness. In E. Sözer (Ed.), *Text connexity, text coherence: Aspects, methods, results* (pp. 189–218). Hamburg: Buske.

Redish, J. C., & Selzer, J. (1985). The place of readability formulas in technical communication. *Technical Communication, 32*(4), 46–52.

Rieger, C. (1975). Conceptual memory. In R. C. Shank (Ed.), *Conceptual information processing.* Amsterdam: North-Holland.

Riley, K. (1987, December). *Language theory: Application vs. practice.* Paper presented at the MLA Convention, New York City.

Riley, K. (1988a). Conversational implicature and unstated meaning in professional communication. *Technical Writing Teacher, 15*(2), 94–104.

Riley, K. (1988b). Speech act theory and degrees of directness in professional writing. *Technical Writing Teacher, 15*(1), 1–29.

Riley, K. (1993). Telling more than the truth: Implicature, speech acts, and ethics in professional communication. *Journal of Business Ethics, 12,* 179–196.

Rude, C. D. (1991). *Technical editing.* Belmont, CA: Wadsworth.

Schriver, K. A. (1989a). Document design from 1980 to 1989: Challenges that remain. *Technical Communication, 36*(4), 316–331.

Schriver, K. A. (1989b). Evaluating text quality: The continuum from text-focused to reader-focused methods. *IEEE Transactions on Professional Communication, 32*(4), 238–255.

Schriver, K. A. (Ed.). (1989c). Document design moves into the next decade. *Technical Communication, 36*(4).

Schriver, K. A. (1993). Quality in document design: Issues and controversies. *Technical Communication, 40*(2), 239–257.

Searle, J. R. (1969). *Speech acts.* Cambridge, UK: Cambridge University Press.

Searle, J. R. (1975). Indirect speech acts. In P. Cole & J. Morgan (Eds.), *Syntax and Semantics* (Vol. 3, pp. 59–82). New York: Academic Press.

Searle, J. R. (1976). A classification of illocutionary acts. *Language in Society, 5,* 1–23.

Sidner, C. L. (1983). Focusing in the comprehension of definite anaphora. In M. Brady & R. C. Berwick (Eds.), *Computational models of discourse* (pp. 267–330). Cambridge, MA: MIT Press.

Smith, J. M. (1992). *SGML and related standards: Document description and processing languages.* New York, NY: Ellis Horwood.

Stotsky, S. (1983). Types of lexical cohesion in expository writing: Implications for developing the vocabulary of academic discourse. *Research in the Teaching of English, 34*(4), 430–446.

Thompson, I. (1985). The given-new contract and cohesion: Some suggestions for classroom practice. *Journal of Technical Writing and Communication, 10*(3), 205–215.

Tierney, R. J., & Mosenthal, J. H. (1983). Cohesion and textual coherence. *Research in the Teaching of English, 17*(3), 215–229.

Tzeng, O. J. L., & Singer, H. (Eds.). (1981). *Perception of print: Reading research in experimental psychology.* Hillsdale, NJ: Lawrence Erlbaum Associates.

van Dijk, T. A. (1980). *Macrostructures: An interdisciplinary study of global structures in discourse, interaction, and cognition.* Hillsdale, NJ: Lawrence Erlbaum Associates.

Vande Kopple, W. J. (1982). Functional sentence perspective, composition, and reading. *College Composition and Communication, 33*(1), 50–63.

Walter, M. (1992). A look inside J-CALS: Integrated approach to technical manuals. *Seybold report on publishing systems, 21* (pp. 13–17). Media, PA: Seybold Publications.

Wertheimer, M. (1938). Laws of organization in perceptual forms. In W. D. Ellis (Ed.), *A source book of Gestalt psychology* (pp. 71–88). London: Routledge & Kegan Paul.

Winograd, T. (1972). *Understanding natural language.* New York: Academic Press.

Witte, S. P., & Faigley, L. (1981). Coherence, cohesion, and writing quality. *College Composition and Communication, 32*(3), 189–204.

Ziff, P. (1984). *Epistemic analysis: A coherence theory of knowledge.* Dordrecht, Holland: D. Reidel.

Author Index

Subject Index

A

Accuracy, *see* Coherence principles, quantity

Adequacy, *see* Coherence principles, quantity

Alliteration, *see* Cohesion elements, auditory/
phonological

Ambiguity, 28, *see also* Coherence principles,
manner, 27–31

Anaphora, *see* Cohesive elements, semantic,
reference

Application, *see* Levels of inquiry

Applied linguistics, *see* Levels of inquiry

Audience analysis, *see* Document design,
implications of theory for practice and
pedagogy

B

Bridging
and accuracy, 33
and adequacy, 31
and clarity, 28
definition of, 17–18

C

Case roles, *see* Discourse elements, *see also*
Cohesive elements, semantic, 47–51,
55–56

Clarity, *see* Coherence principles, manner

Cognitive foundation for document design,
1–3, 7, 9–11

Coherence
definitions of, 5, 12
gradability of, 18–19, 29, 33, 81
variability of, 19–22, 29, 33, 37–38, 81
limitations on, 20, 38

Coherence principles, *see also* Preference
conditions, 81–83, 89
categories of, 16–35
manner, 27–31
quality, 33–35
quantity, 31–33
relation/continuity, 16–27
and cohesion, 8–9, 77–83
meta-principles, 57–64
summary of, 74

Coherence research, 5–9
in linguistics, 5–7
in psychology, 7–9

Coherence and recipient knowledge, 22–27,
see also Recipient of discourse, types
of knowledge, 22–27

Cohesion
definitions of, 5–6, 12
gradability of, 40, 51–52

D

E

F